CIRCLES OF REFLECTION

TO DRS. THOMAS AND MARTHA CARTER,
LONGTIME FRIENDS, SUPPORTERS, AND
PATRONS OF THE CLEVELAND MUSEUM OF ART

Circles of Reflection

THE CARTER COLLECTION OF CHINESE BRONZE MIRRORS

Ju-hsi Chou

Technical analysis by Bruce Christman

THE CLEVELAND MUSEUM OF ART

Board of Trustees

Published on the occasion of the exhibition *Circles of Reflection: The Carter Collection of Chinese Bronze Mirrors*, The Cleveland Museum of Art, 17 September–26 November 2000.

Published by
The Cleveland Museum of Art
11150 East Boulevard
Cleveland, Ohio 44106

Library of Congress card number: 00–107401
ISBN: 0–940717–59–X

Editing: Kathleen Mills
Design: Laurence Channing
Production: Charles Szabla and Carolyn K. Lewis
Photography: Gary Kirchenbauer
Typographic systems consultant: Margo Frey
Chinese typesetting: Technical Translation Services, Inc.
Printing: Watt Printing Inc.
Composed in Monotype Walbaum
Front cover: *Mirror with Deities and Animals Surrounded by Rings of Squares and Semicircles*, cat. no. 38. 1995.333
Back cover: *Mirror with a Coiling Dragon*, cat. no. 72. 1995.367

Contents

Foreword vi
Katharine Lee Reid

Acknowledgments vii

Chronology ix

Map of Production Sites x

Introduction 1
Ju-hsi Chou

Plates 15

Catalogue of the Exhibition

Phase I: Warring States Period 23

Phase II: Han and Post-Han Dynasties 31

Phase III: Sui and Tang Dynasties 59

Phase IV: Liao, Song, and Jin Dynasties 81

Phase V: Yuan, Ming, and Qing Dynasties 91

Nos. 1–16 *Ju-hsi Chou and Christine Starkman*
Nos. 17–92 *Ju-hsi Chou*

Appendix: Technical Issues

Making the Mirrors 98
Bruce Christman

List of the Carter Collection 111

Bibliography 129

Foreword

It is to the great benefit of the Cleveland Museum of Art that Thomas and Martha Carter of Madison, Wisconsin, have generously donated their collection of ancient bronze mirrors. Dr. Thomas Carter is a physician. Dr. Martha Carter is a native Clevelander who has always regarded the Cleveland Museum of Art as "her" museum. She grew up in University Circle and during her early years worked in various capacities for this museum. She was also closely connected to our former director and my father, Sherman E. Lee, and received her Ph.D. at Case Western Reserve University under his supervision. Although her own field is Near Eastern art, she remains, as a result of Dr. Lee's influence, very much enamored with Chinese art. This mirror collection, the majority of which is Chinese in origin, was formed under her direction, with encouragement and unstinting support from her husband, Thomas.

Five years ago, the bulk of their collection was given to us in honor of my father. At the time, the curator in charge of Chinese art was Keith Wilson, who subsequently departed for the Los Angeles County Museum of Art. Ju-hsi Chou was appointed to fill the position left vacant. On his arrival in January 1998, Dr. Chou assumed responsibility for mounting an exhibition and writing the accompanying catalogue. His efforts are especially appreciated because as a painting specialist, he had never before ventured into the subject of the bronze mirror. Bruce Christman, chief conservator, was an enthusiastic collaborator from the very beginning of the project, and approached the topic from a technical and metallurgical angle. Christine Starkman is responsible for some early research that resulted in a number of catalogue entries.

As the museum's new director, I represent the entire institution when I express my gratitude to the Carters. With their donation, the Cleveland Museum of Art now has one of the finest public collections of Chinese bronze mirrors in the United States. The Carters' passion for the subject of this exhibition and for sharing it with a broad public is evident not only in their generous donation, but also in their enthusiasm for mounting a symposium to accompany the exhibition. We were honored and grateful to secure financial support for the symposium from the Chiang Ching-kuo Foundation for International Scholarly Exchange (USA). In all, the rich interest this refined topic has inspired from every quarter exemplifies the rewards of the museum enterprise at every scale. I especially admire the cooperative spirit that prevails in our museum, where the curatorial, conservation, and education staffs and designers have all worked together to bring this project—the exhibition, the catalogue, and the symposium—to fruition.

Katharine Lee Reid
Director

Acknowledgments

BY CLEVELAND MUSEUM OF ART standards, *Circles of Reflection: The Carter Collection of Chinese Bronze Mirrors* is a modest project. And it has remained modest in scale, even though requiring cooperation from many individuals.

I must first acknowledge the invaluable help from our colleagues in other museums and institutions, who have either pointed me in the right direction for mirror research or provided me with the opportunity to view Chinese bronze mirrors in their collections: Claudia Brown, Phoenix Art Museum and Arizona State University; Terese Tse Bartholomew, Asian Art Museum of San Francisco; Jing Gao and Tung Wu, Museum of Fine Arts, Boston; Robert Mowry and Melissa Moy, Harvard University Art Museums, Cambridge; Jenny So, Freer Gallery of Art/Arthur M. Sackler Gallery, Washington, D.C.; Zhixin Jason Sun and Hwai-ling Yeh, Metropolitan Museum of Art, New York; Christina Tang, Hong Kong Museum of Art; and Anita Wong, University Museum and Art Gallery, University of Hong Kong.

Within the Cleveland Museum of Art, the support of the late Director Robert P. Bergman, former Acting Director Kate Sellers, and Chief Curator Diane De Grazia must be acknowledged. Our new director, Katharine Lee Reid, continues the tradition and has lent a helping hand. Chief Conservator Bruce Christman has been a willing participant in this project from its inception. He has brought a scientist's eye and an approach that complements well my own art historical viewpoint.

In research, I was also fortunate in enlisting help from Yunah Sung of the museum's Ingalls Library, who has unfailingly provided me with bibliographical assistance and has also secured critical titles in a prompt and efficient manner. In the compilation of research files, help came from an unexpected quarter. Chen (Nikki) Yee-Hsiun, an intern from the Tainan National College of the Arts, arrived in time to join us on the research team. Within two months, Miss Chen delved into archaeological reports with focus and intensity and came up with a systematic compilation, without which my progress would have been seriously stymied.

The excellent photographs used in this catalogue are the work of Gary Kirchenbauer, and their production the result of the entire photography department. The superb design for the catalogue originated in our publications department, led by Laurence Channing. To Kathleen Mills I extend thanks for her professionalism, insightful editing, and helpful guidance. The fine installation of the exhibition can be credited to our design team, with Jeffrey Baxter contributing his skill and providing input along the way. Katie Solender, exhibitions manager, also rendered us a great service by making the labels accessible to a broad audience. I also wish to thank the education department's Joellen DeOreo and Jeremy Shubrook for their assistance in organizing the symposium. Not to be forgotten is Rob Krulak in the development office, who made a special effort to search for outside grants. Finally I must also pay tribute to the registrar's office, as its support—as invisible as it is real—has been critical to the project's success.

As usual, the staff in the Asian department has been the backbone of this endeavor. Nancy Grossman, curatorial assistant, and Beth Sanders-blevans, administrative assistant, have lent their hands to an assortment of tasks relative to the exhibition, and in particular were willing to read and correct the catalogue manuscript prior to the final version. In particular, Beth spent countless hours in going through the entire manuscript and made invaluable suggestions that helped to shape the mode of presentation to a level that had been previously lacking. Indeed, their dedication and cooperation played a significant role in the success of the catalogue and the exhibition. To them I owe a profound gratitude.

To the Chiang Ching-kuo Foundation for International Scholarly Exchange (USA), I wish to express my sincere gratitude for its conference grant. These monies enabled us to arrange a symposium where scholars in mirror studies from Asia, Europe, and the United States can gather together to exchange data and ideas. Above all, I would like to thank Drs. Thomas and Martha Carter for their generous donation, which inspired this project, and for their unbounded enthusiasm for teaching others about this topic. Without their excellent collection to inspire me, the task of learning a new subject, at my current stage of career, would have been unthinkable.

Ju-hsi Chou
Curator of Chinese Art

A Note on Romanization

Romanization of Chinese characters in this catalogue will adhere to the *pinyin* system, with the following exceptions: terms and names in titles of publications using different systems of romanization; self-chosen names of modern Chinese scholars in the West, familiar through previous publications, e.g., Cheng Te-kun instead of Zheng Dekun; place names outside the People's Republic of China, the usage of which is accepted internationally, e.g., Hong Kong (not Xianggang) and Taipei (not Taibei).

Relevant Dynastic Chronology

Warring States	475 – 221 BC
Qin	221 – 207 BC
Han	206 BC – AD 220
Western Han	206 BC – AD 8
Xin	9 – 24
Eastern Han	25 – 220
Three Kingdoms	220 – 280
Shu	221 – 263
Wei	220 – 265
Wu	222 – 280
Jin	265 – 420
Western Jin	265 – 316
Eastern Jin	317 – 420
Northern and Southern Dynasties	420 – 589
Sui	581 – 618
Tang	618 – 907
Five Dynasties	907 – 960
Liao	916 – 1125
Song	960 – 1279
Jin	1115 – 1234
Yuan	1279 – 1368
Ming	1368 – 1644
Qing	1644 – 1911

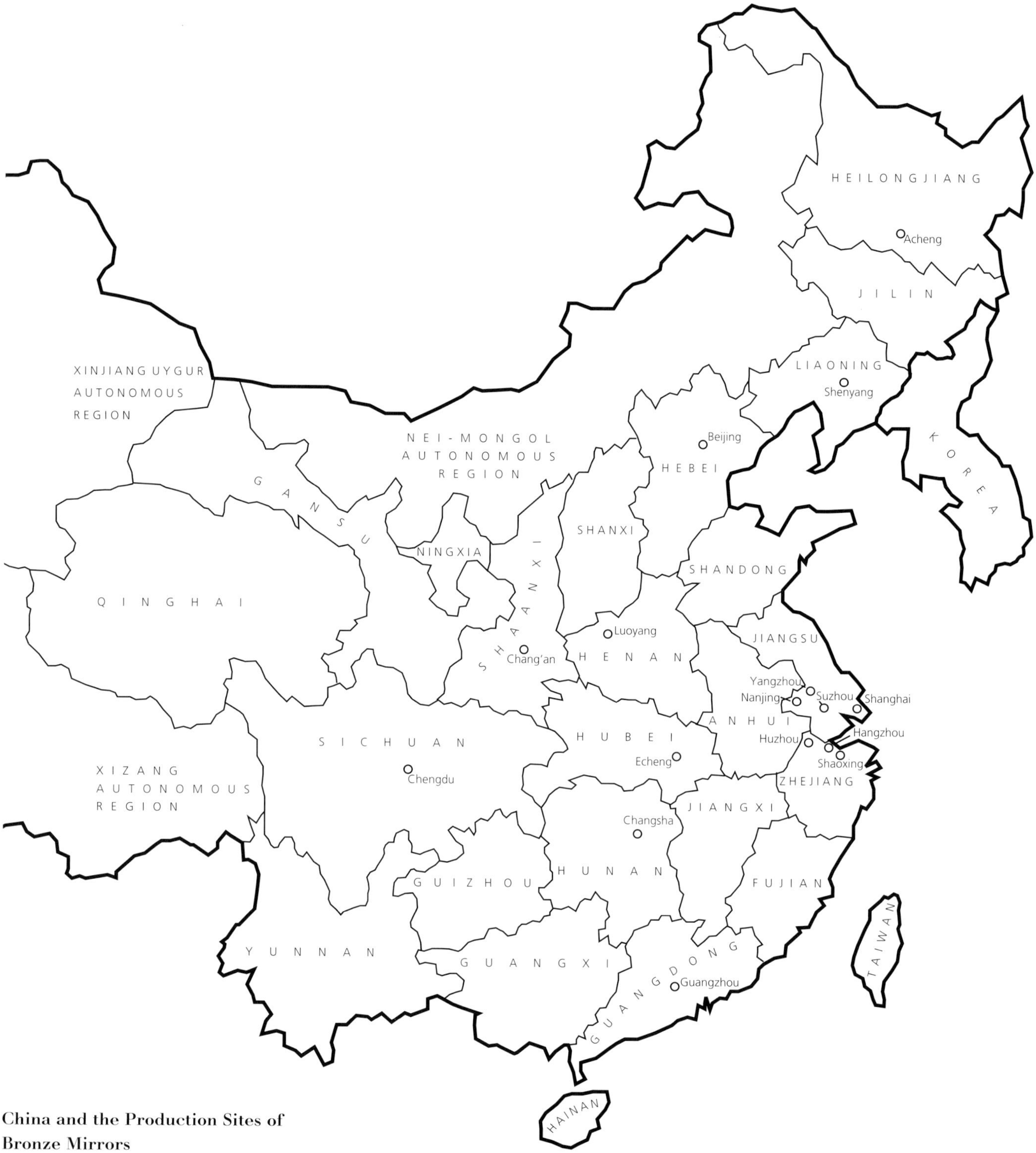

China and the Production Sites of Bronze Mirrors

A number of sites have been identified as centers for mirror production. To satisfy the demands of the court and court officials, governmental foundries were located either in the capitals, secondary capitals, or in their vicinity. Examples are Chang'an and Luoyang during the Han and Tang dynasties. Commercial centers also took advantage of their proximity to copper mines, as indicated by the urban development around them as well as their strong market demands. Shaoxing, Yangzhou, and especially Huzhou, are good examples.

Introduction

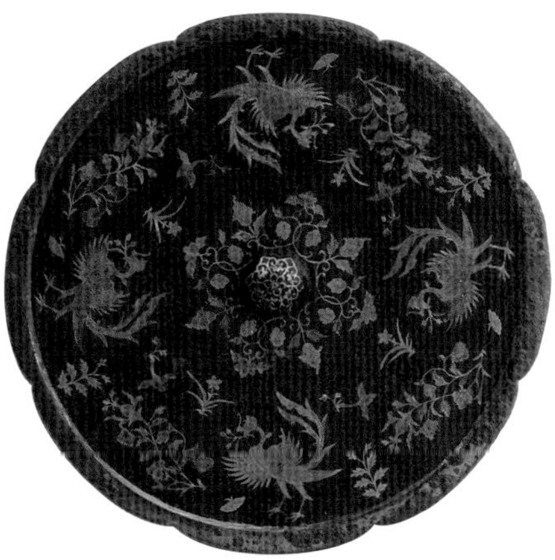

Fig. 1. *Mirror.* Bronze with silver and gold cutouts over a lacquer base. Tang dynasty, 8th century. Diameter 19.2 cm. Leonard C. Hanna Jr. Fund 1973.74

THE CLEVELAND MUSEUM OF ART is gratified by the donation of a collection of bronze mirrors from Drs. Thomas and Martha Carter in honor of CMA's former director, Dr. Sherman E. Lee. That the museum has been acquiring Chinese bronze mirrors from as early as 1915 is well known. Prior to the Carter donation, there had been a small number of mirrors dating to the Warring States, Han, and Tang periods.[1] This eventually culminated in the acquisition, under Dr. Lee, of a marvelous Tang mirror whose back is decorated by cut pieces of gold and silver in floral and avian shapes over a lacquer base, a technique known as *jinyin pingtuo* 金銀平脱 (the gold and silver pattern leveling with and emerging [from a lacquer base]; fig. 1). Even today such examples are quite rare, and those archaeologically excavated are even more so.

THE CARTER COLLECTION

The Carters' donation of bronze mirrors, which arrived at the museum in two separate installments in 1995 and 1999, further augmented and expanded this base. Numbering 143 in total (with two mirror stands, six Japanese mirrors, and two Near Eastern mirrors in addition to the Chinese mirrors),[2] this collection represents a quantitative as well as a qualitative leap and places the Cleveland Museum of Art above and beyond most, if not all, existing mirror collections in other U.S. public institutions. The quantitative side is self-apparent. In the matter of quality and substance, even a casual glimpse will reveal a number of surpassing works, where refined casting, lustrous surface, and significant imagery combine to achieve a rarefied state of visual and conceptual eloquence. For instance, the small *Mirror with Four T's* (no. 2) from the Warring States, with a patination known as *heiqigu* 黑漆古 (black lacquer antique), offers a design of immense potential. Several of the TLV mirrors in the Carter collection (nos. 26, 28, 29), exquisitely cast, are comparable to the best known examples in any collection, just as mirrors of the later Han provide evidence of fine imagery and extraordinary detailing (nos. 23, 38). Of the Tang mirrors, the animal-and-grape design (nos. 52–56) on the mirror backs typifies the period's lush exuberance, while other designs also attain artistic height by combining their speculum metal with a plethora of zoomorphic, avian, and vegetal/floral images. Turning to later epochs, who could ignore the Song mirror's subdued and understated charm (no. 77), especially set against the primal power and vigor of the Jin (no. 86) and Yuan pieces (no. 87)? Nor can one overlook the simple casting of an essentially unadorned Huzhou mirror (no. 90)—Huzhou being a center of mirror production that thrived during the Song period and lasted until the eighteenth century—one that derives its chief appeal from its grayish blue, steely surface.

If the CMA collection of mirrors in earlier years tended to concentrate on the Tang and pre-Tang periods, that was due to the taste of the time. In view of the general state of mirror studies from 1915 all the way to the 1970s or early 1980s, this is not at all surprising. By contrast, the scope of the Carter collection is far more inclusive. To complement the splendid mirrors of earlier periods, the Carters have added the post-Tang specimens (nos. 76–92). Since the post-Tang mirrors have only been given recognition in very recent times—and primarily in China—the Carters' willingness to place a portion of their resources in these less established works was truly a daring act, and one that should command our admiration.

Chronologically, the full range of the Carter collection stretches between the Warring States period and the Qing dynasty, a span of roughly 2,500 years. Statistically, a breakdown by period of the Carter bronze mirrors is as follows: Warring States (475–221 BC), 13; Han to the Northern and Southern dynasties (206 BC–AD 589), 55; Sui and Tang dynasties (AD 581–907), 37; post-Tang, 27. When viewed in this light, the Carter collection must be thought to be unique in the modern West, a worthy successor to the Todd collection of the 1930s.[3] It should be noted that the Todd collection was formed during an age that was innocently indiscriminating. In turn, the Carter collection represents the last decades of the twentieth century, when knowledgeable connoisseurs were responding to a full range of recent archaeological and scholarly appraisals, and made a bold attempt toward a broader base.

Because of this, the Carter collection of bronze mirrors embraces a variety of production techniques, alloy compositions, and patination. Thus we can find, in the mirrors themselves, a range of casting methods, from simple mold casting (nos. 1–45) to *cire perdue* (no. 46 onward), from the single, unitary disc to the double-tiered approach (no. 4), and on to the sophisticated technique previously mentioned of *jinyin pingtuo* (no. 61). In alloy composition, a large portion of the Carter collection reveals a high-tin, low-lead combination (e.g., the pre-Tang and Tang specimens as well as some of the Song and Jin), while others are of a reduced-tin, higher-lead variation. If the high-tin, low-lead combination led to fine casting, precise detail, and speculum surface, the disadvantage was that it also caused the mirrors to become hard and brittle. The latter formulae enabled a lower melting temperature and a greater ease of flow in the molten state, attributes that are advantageous during production, but with a concomitant loss of accuracy in detail due to greater shrinkage when the bronze turned from the molten to the solid state.[4] The collectors were also sufficiently attentive to the *current* conditions of these bronze mirrors—as opposed to the time of casting. These pertain to varying states of preservation: notably the lustrous black of the famed *heiqigu*, for which there are a number of fine specimens (nos. 2, 34, 53, 66a, 72), the silvery metal of *shuiyinqin* 水銀沁 (nos. 50, 55, 56, 59, 66b), the water patination of olive green (no. 80), and the dark brown shades of the high-lead, low-tin alloy of the Song and Jin periods (nos. 82, 86). From the Qing period, we can attest to the yellow tone of the brass in the mirror with a handle (no. 92). Its inclusion of a handle can be considered a belated revelation of Western impact.[5]

Then there is, among the Carter mirrors, the varied spectrum of sizes, types, approaches, viewing orientations, and wide-ranging display of themes and symbolism in the decorations. (These decorations, we hasten to add, appear only on the mirror backs, not on the reflective surfaces.) As far as size and dimension are concerned, the Carter collection contains miniature "pocket" mirrors (nos. 48, 54) as well as large Tang and Song pieces (nos. 68, 79), with diameters exceeding 28 and 35 cm, respectively.[6] In shapes and typology, the collection embraces a considerable scope: the circular tends to dominate, of course, this being the Sinic tradition of mirrors (see below). There are also the lobed (*kuihua jing* 葵花鏡; nos. 58, 59, 65–67, etc.), the foliated (*linghua jing* 菱花鏡; nos. 62–64, 68, 80–82), the cruciform (no. 78), and the square contour (no. 77). (The only shapes missing from this collection are the Huzhou mirrors produced during the Song dynasty, where they took on a number of variations that seem to be far-fetched to the mirror form, but aligned with other antique objects cherished during the period.) In decorative modes, the approach varies from the earlier, dual structure of isolated images over a ground maze (nos. 1–3, 5–9), versus its dissolution in favor of a unified treatment (no. 10 onward), and the eventual progression from the subtly differentiated layers of planes (nos. 17, 34) to higher relief (nos. 38, 39, 51, 53, 55, 56). In regard to viewing orientation, the circular disc of the bronze mirror favors revolution around the central knob—in either centrifugal or centripetal mode—to one of the following: the division of the circle into zones of varying viewpoints (no. 41), and, far more enduringly, the pictorial mode with the stress on the vertical axis (nos. 66a–b, 67–71).

In terms of themes and subjects, the range is immense. Included here are the divine couple of Dong Wanggong 東王公 and Xi Wangmu 西王母 (nos. 38, 39, 41), and the Four Spirits of Blue Dragon, Red Bird, White Tiger, and the Dark Warrior of intertwining turtle/snake (no. 49). Mythical and historical personages include Fuxi 伏羲, Nuwa 女媧, and Huangdi 黃帝, the Yellow Emperor, along with the implicit or actual presence of his nemesis, Zhiyou 蚩尤 (no. 38, in the narrow band of decoration along the border); the infinitely wise and sagacious Yao 堯 and Shun 舜 (no. 41); the rise and fall of Fucha 夫差, the Lord of Wu, and Goujian 句踐, the ruler of Yue (no. 32); and Bo Ya 伯牙 the musician and his friend, the music connoisseur Zhong Ziqi 鐘子期 (no. 38). With respect to folktales, it is the legend of the Herdboy and the Weaver Maid (no. 81) that vies with the fabled Emperor Ming's visit to the moon (no. 83); the immortals crossing the sea (no. 82) can be pitted against a long-forgotten hero, Ci Fei, who single-handedly combated the evil dragons and saved the shipful of lives (no. 80). In terms of zoomorphic images, the dragons and the phoenix, the fabulous *qilin* and *pixie*, are just as favored as other creatures. At times, even simple insects and birds can make up a considerable repertoire (no. 60, for example), not to mention the world of plants, flowers, and fruits, including the grape (nos. 52–56), the lotus (nos. 87, 88), the peony and camellia (no. 59), and so on.

The world of the mirror is also the mirror of the world. What is reflected in front of the mirror may be illusory, but those who come and look into its reflective surface will never be lonely when accompanied by these "submerged" images on its back.

THE MIRROR AS ART FORM: ITS INHERITANCE AND INDEPENDENCE

If some of these themes sound familiar—at least to scholars of Chinese art—they are indeed. Not a few are also present in bas-reliefs, paintings, textiles, and other art forms, the total of which, of course, encompasses a far broader scope. The Dong Wanggong and Xi Wangmu pair, Fuxi and Nuwa, and the Four Spirits are an integral part of the Han repertory of tomb decoration.[7] The forgotten hero of Ci Fei (no. 80), for instance, was originally the subject of a Tang painting which attracted the attention of Li Bai 李白 (701–762), the great Tang poet, and encouraged him to sing the hero's praises.[8] Along this same line, an innocuous design (no. 62) may be linked, however indirectly, to the illustrious artistic family of the Yan Lide 閻立德 (d. 656) and Yan Liben 閻立本 (d. 673).[9] The zoomorphic and vegetal motifs were not only for mirrors. They are also observed on textiles during the Han and the Tang periods. Today, even the marvelous Warring States mirrors with T-shape decoration (no. 2) are believed to have their design traceable to an early weaving pattern.[10]

But above all, we must also admit that the TLV mirror, around which so much has been written, began its life by superimposing, on its circular bronze disc, the game board pattern of the *liubo* 六博.[11] After Western scholars had initially lavished attention on its cosmological significance, that realization in itself was a major turnaround.[12] This was accomplished in recent years through examination of actual finds of *liubo* paraphernalia and also from evidence derived directly from the bronze mirrors themselves. Of course, the pattern of a game board can also harbor, or be made to gain, such symbolism.[13] But who can deny the game board derivation when it was a mirrorsmith who admitted as much in an inscription on his own work? As he said:

In this era of Han, fine copper is being mined at Danyang.
Mixed with silver and tin, the alloy turns clear and luminous.
Engrave the liubo *pattern with a square in the middle.*
Commanding the four directions are the Dragon at left and Tiger at right.
The Red Bird and the Dark Warrior are in tune with the forces of yin *and* yang. . . .[14]

The point is, the mirror as an art form was never a closed circuit. Rather, it was open to a multitude of sources and influences. From its very inception, the mirror as an art form was the beneficiary of a long history of bronze casting, which had begun sometime in pre-Shang eras (before the sixteenth century BC). If the metalsmith once had set his sight on the making of magnificent ritual vessels, on the complex process of forging keen-edged weapons, and on sundry artifacts, then there was nothing to stop him from venturing into mirror production, a task which was, by comparison, far simpler in technical terms. The fact is, from the Warring States period onward, the mirrormakers appeared to have shared a ready knowledge of the best alloy formulae which could induce a luminous, reflective surface, testifying to the inheritance of accumulated experience. In the search for such a surface, the art of the mirror defined its own particular path for an extended period, until the realization, after the eleventh century AD, that it only required a reflective coating in order to serve its function. The base mirror itself did not require such a rigid standard of alloys as previously held.[15] Only then did the nature of the alloy change remarkably, with a reduction in tin and an increased percentage of lead, and the accompanying advantages as well as disadvantages.

Along a similar line, the mold casting and *cire perdue* were practices that were readily transported from the casting of ritual vessels and implements to the world of mirrors. Together with this transference of medium, the transfusion of design elements took place as well. The feathered hook, the serpentine interlace, the quatrefoil, the "thunder" pattern and the whorls, etc., were to a large extent drawn from a pool of motifs evolved in the broader spectrum of bronze art. Indeed, if a metalsmith could take a horse and chariot fitting and turn it into a double-tiered mirror, there is no reason to bar him from venturing further (no. 4).[16] Other inspiration—from bas-relief, painting, textile, game board, and so on—diversified and amplified the shared pool of decoration. For those attentive to inscriptions, the content or messages were not only socially, politically, or religiously compelling, but could also be related to the art of poetry and that of calligraphy.[17] In the latter aspect, a range of written scripts can be observed—from seal script through the clerical to the regular script—which adds critical and otherwise unavailable data in support of the evolution of China's written language as well as styles of its execution.

Yet, amidst all the impacts and absorption thereof, mirror art in China can still stand as a unique instrument which possesses an identity all its own. Just as in the TLV mirror, the *liubo* board was but a start, not the end; it evolved and metamorphosed beyond the original perimeter (nos. 24–29). Against the pool of shared myth and themes, the unshared are perhaps just as substantial and conceivably wider in scope. Why, if we were to ask a rhetorical question, was there not a single illustration, in Han mirrors, of the assassin-hero Jing Ke and the King of Qin, the future First Emperor of China, a subject widely known in reliefs from Shandong to Sichuan, virtually across China?[18] Inherent characteristics of the mirror, such as its limited size and shape, would have in any case precluded anything that required lengthy exposition, just as its proximity to the user and its intimate function[19] would have favored an overall clarity of detail and encouraged a positive and benevolent spirit.[20] The opposite would have had enormous difficulties in ever making it to the mirror's domain; we refer to such tendencies as a grand gesture, a monumental undertaking, and, in particular, the celebration of acts of violence.[21] Instead of being a merely passive recipient of sources and resources, the mirror, through the able mind and the hand of its maker, metaphorically speaking, carved out its perimeter and actively selected those elements which were agreeable and rejected those that were not.

IN THE SEMBLANCE OF SUN AND MOON

Of all the attributes of the Sinic bronze mirror, including those already enumerated, the most important is still its circularity. From archaeological evidence—even counting what we now know about mirrors from the Qijia culture (ca. 2500–1500 BC), when mirrors made their first appearance, and from the Shang (ca. 1600–1100 BC) and the Western Zhou (ca. 1100–771 BC)—the Chinese bronze mirror started out as a bronze disc with a central knob at the back, the perforation of which allowed insertion of a ribbon or tassel for handling purposes.[22] It differs from its Western counterpart in the absence of a handle, although at a later date it was clearly affected by the convenience of that particular element (no. 92). Other contours or shapes existed, as partially detailed above. Even if we only had the Carter collection as our reference, it is apparent that the Sinic tradition of the mirror lies in its close adherence to the circular form.

This was so from the very start and has continued for more than two millennia. The circle is the mirror, and the mirror the circle. Within its contour, the mirrormaker could either arrange the circle into concentric series or make use of its variants, such as smaller revolving circles, nipples, or linked arcs—those which are either the whole or the parts of a circle, or which divide it into segments and zones. Indeed, by utilizing the linked arcs in a concentric manner, the craftsman could fashion the image of the sunburst (nos. 14–17). So said a mirrorsmith (no. 14):

By the light of the sun, the world is made bright!

At a later time but under similar circumstances, another mirrorsmith made known his metaphoric parallel (no. 15):

Its inner purity is shown in perfect illumination.
Its light is the image of the sun and moon. . . .

If among Chinese scholars mirrors bearing the former inscription are known as the *riguang jing* 日光鏡 and those with the second inscription are known as *zhaoming jing* 昭明鏡, the distinction lies in their respective utterance as in the inscriptions, not in their design. It is the sunburst that provides the basic image for both "types"—and more. In the image of the sun and moon, the mirror proclaimed, and so the sunburst lasted through the four hundred years of the Han era. The sunburst design also applies to the so-called "copper essence" mirror (*tonghua jing* 銅華; no. 16), or the "mirror of linked arcs" (*lianhuwen jing* 連弧紋鏡; no. 17). And in a radical transformation, the sunburst image also turns itself into the so-called "mirror with quatrefoil and transformed images" (*bianxing siye wen jing* 變形四葉紋鏡), a potentially "psychedelic" image of throbbing forces and rhythms (no. 37).[23] In this context, we might suggest that the current typological classification is far too rigid, as it neglects to observe the linkages and fluid changes among types.[24] In too many instances, mirror studies become an enumeration of types without a keener understanding of the complex process—artistic, psychic, and cultural—in the substratum.

In short, we may deem the history of mirrors in China as either the permutation of the circle or circular movements, whether it is in concentric series (nos. 19–23), revolving images (nos. 38–40, 60, 63, 64), more circles around circles (nos. 22, 23), or, intriguingly, contrary approaches that to a limited extent counter the mirror's circularity. We have already mentioned the division of the circle into zones, with varying viewpoints (no. 41), or pictorial modes, which sustain a vertical axis as the viewing angle (e.g., nos. 67–71). We can further mention a decked hierarchy—reserved generally for immortals—as another alternative, even though it is not represented in the Carter collection.[25] Still, the circle is the base, within which the mirrorsmith either struggled in terms of it or, within limits, moved against it.

One of the most powerful devices in this struggle consists of the introduction of the square within the sphere of the circle (nos. 2, 10, 11). The potential in this is unlimited, for, within the Chinese tradition, this square within the circle not only exemplified the duality of the struggle, but also depicted the bipolarity of heaven and earth ("heaven is round and earth square"), or to put it in another way, it gave a palpable sense of the earthly domain and spiritual realm. In this bifurcated sphere, the Four Spirits and the host of other images, such as *yuren* 羽人 (feathered immortals) and benevolent animals are brought into being. Inasmuch as the Four Spirits command the cardinal directions, seasons, elements, or forces, as well as a spectrum of outer and inner virtues, this is, in the context of the TLV typology, the mirror of the cosmogram (nos. 26–29). In front of this mirror, the owner of the reflected human image was but an integral part and can be posited within the cosmogram to form a holistic whole. It was this individual who, whether male or female, aspired toward progeny, longevity, power and wealth, peace on earth, a desire for immortality, and the urges toward the celestial planes. To wit (no. 28):

A fine mirror it is, my making of it!
I can show the ageless immortals up above.
In thirst, they drink from the jade spring,
When hungry, they partake of jujubes.
They wander over the earth and belittle the Four Oceans.
They live as long as metal and stone,
And are the true protectors of our kingdom.

However humble it was at the beginning, the TLV mirror became *the* dominant type and inspired prolific production during the whole of the Han period and longer. A number of the later mirror types descended from it—for example, those that retained the centralized square, sans the T's, L's, and V's, but with a selection of quartets of animals or deities (nos. 30, 31) or, in the Sui-Tang period, the Four Spirits revolving around it (no. 49). Other mirrors, whether of the same branch or stemming from other typologies, managed to reject the square altogether, at least for a limited duration.[26] Other approaches also existed, which comprised those mirrors with rings of squares and semicircles (nos. 38–41), and those that made use of figural narratives and iconic presentations (nos. 32, 38, 41). Meanwhile, mirrors with multiple concentric circles remained in force (nos. 23, 47), and those with "psychedelic" images continued to pulsate and throb, as leaves turned into bats, in combination with phoenix and/or animal masks (nos. 36, 37).

With this discussion we are but recounting some of the permutations that occurred at the decline of the TLV typology, when mirrorsmiths were compelled to fill in the vacuum with other, more viable alternatives. Every change signals an attempt to break through the confine of simple circularity and represents an urge to create new frameworks or vocabularies. Some were just as successful as the TLV mirror, and others were less so and were shortly aborted (no. 41).

During the Sui-Tang period, the mirrorsmith continued to work with the circle as well as to produce modifications. In the former case, the animal-and-grape typology is perhaps the most salient (nos. 52–56). The attempt to modify the outer contour of the mirror resulted, among other mirror types, in the lobed mirror (*kuihua jing;* nos. 58, 59) and the foliated mirror (*linghua jing;* nos. 62–64). Granted, these contours remain as centralized as their parental image (the circle), thus their modification was not so much revolutionary as a minor adjustment. Drawing parallels by means of the luminous reflection of the mirror's surface, this metaphorical link to the sun and moon remained steadfast:

照日菱花出，臨池滿月生。

The sun shines on the foliated shape.
A full moon rises over the pond.[27]

By and large, in the mind of the Tang Chinese, the more apt comparison is between the mirror and the moon. Unlike the brilliance of the sun, which produces light, both the mirror and the moon are but reflective. Even though this particular piece of scientific knowledge was not known at the time, the Tang Chinese nonetheless marked their observation by noting, between the mirror and the moon, a luminosity that was not so much overpowering as quietly limpid. Thus the maker of the Tang mirror (no. 51) in the Carter collection speaks to this effect:

鍊形神冶，瑩質良工。
如珠出匣，似月停空。

The spirit merges with the forms that are being wrought.
The lustrous metal matches the exquisite craftsmanship.
Like a pearl, it emerges from the container box—
A virtual moon suspended in the firmament.

This emphasis coincides with the emergence of a new sentiment, which luxuriated in the newfound literacy and poetry as well as a sensual tone that far surpassed the old well wishes for longevity, high posts, wealth and progeny, even for immortality. Here the mirror is compared to the incandescent pearl and the cool light of the moon, and within that context, the femininity of the mirror ascended, and a subtle, erotic connotation was made clear:

當眉寫翠，對臉傳紅。
綺窗繡幌，俱含影中。

[It enables] the lady to draw her eyebrows
And apply rouge to her face.
Even those elegant windows and the embroidered curtains:
Can all be reflected within.

During the Tang, it was the famed animal-and-grape mirrors that reached apogee. Indeed, in this type, the sumptuousness of an epoch was made clear: in the inner circle, the animals in high relief assumed various poses sometimes by themselves and at other times hovering over the grapevine; in the outer circle, the clinging vines and clusters of grapes took control, together with a full range of insects and birds. Were these the celebrations of an earthly garden? We think not. It is more likely a celestial one, with the human still looking into it through the reflective surface. That, of course, was but one major, albeit narrowly confined type of mirror in the early Tang period, which remained by and large circular, but with a few square ones as well. Meanwhile, the lobed and foliated specimens expanded beyond the old repertory, while espousing different sets of zoomorphic, vegetal, and human or immortal images, either in revolving movements or stabilized between the dancing phoenixes.[28] In a later stage, the cruciform mirrors, or mirrors with rounded squares, also found a special niche, either to accommodate a Buddhist content (nos. 75, 76) or for floral or other forms of decoration (no. 78).

None, however, can be as radical as the makers of Song mirrors in their willingness to try different approaches. A brief survey would unveil the following shapes among the Huzhou products: peach, shield, vase, bell, the *ding* tripod, etc., all flattened into two-dimensional silhouettes.[29] Whimsical in spirit and antiquarian in derivation, these radical diversions from the circle did not last beyond a limited time. Instead, it was the inheritance of the older types—the circular, the lobed, the foliated, the square, and the cruciform—that remained as the mainstream. The Carter collection also contains a square mirror, with a circle within (no. 77). Subdued in tone, it is nonetheless the height of sophistication and refinement, the very soul of Song art. In the post-Song period, the circular mirror regained its hold, perhaps in recognition of its very "perfection." Even when some mirrors acquired a handle (no. 92), the utility of which was beyond dispute, the design of the circle and that of the handle remained separate. This in itself is a remarkable acknowledgment: that the circle of the mirror is complete in and of itself.

PRODUCTION AND PATRONAGE

A critical question in mirror studies involves the complex relationship between patronage and production. It is reasonable to remind ourselves that bronze mirrors and their usage predicated a level of social and economic prominence that was not always suited to every social class. The poor and the destitute, of course, could not possess such mirrors. Their burial sites, if they had them, would not have survived the ravages of time. Those who could afford bronze mirrors were of the privileged classes—from the lower reaches of officialdom and wealthy individuals to the very center of the imperial court. Court productions during the Han period have survived, and mirrors from the Shangfang 尚方 workshop can still be attested to in fair numbers.[30] By and large, dynastic capitals, or at times secondary capitals, tended to be centers for the production of mirrors. Chang'an, Luoyang, and Echeng (Wuchang) come to mind.[31] On the other hand, regional centers were also active, especially those with copper mines in their vicinity and also with sufficient urban development to ensure a high level of consumption. Suzhou, Shaoxing, and Yangzhou during the Han period and later eras are examples.[32] During the Tang dynasty, Yangzhou produced exquisite mirrors, and sent the best ones to the imperial court as tribute.[33] Even more interesting was the rise and spread of the private workshops alongside the court production centers. Possibly their rise was due either to the limited access to those Shangfang products during the Han period, or perhaps to the ever-widening base of clientele through these centuries, from the aristocrats to the scholar gentry and rising merchant classes, who could afford finer specimens. Indeed, these private workshops vied with one another for market shares or dominance. As the competition heated up, family workshops such as the Wang 王, Chen 陳, Li 李, Liu 劉, Long 龍, and Zhang 張 made themselves known through self-proclamation in their mirror inscriptions. Some of them proudly called themselves masters, *taishi* 太師 or *dashi* 大師. In the Carter collection, two early mirrors bear the marks of these master-craftsmen and deserve mention. Their authors were Master Pan 潘氏 (no. 34) and Master Zhang 張氏 (no. 42). While the former was laconic (merely stating his name on his mirror), the latter was far more verbose, as he made the declaration:

I, Master Zhang, produced this mirror,
When alien tribes in four corners were being subjugated.
Let us praise our Lord, for the people are given respite.
After quelling the Hu and Qiang, he will restore peace on earth.
May you reach the high office and receive your just reward.
May you [gain the power] to protect your parents and descendants from harm.
May your joy be endless.

In fact, the pride taken in the production of mirrors continued into later epochs. During the Song dynasty, Zhou Yuan marked his mirrors with his own name, and eventually announced himself as *the* "famous Zhou Yuan" (no. 82). The Yuan mirror (no. 87) that bears the date of the fourth year of Zhiyuan (1338) also identifies the workshop as that of the Ma family from the district of Hezhong, in present Shanxi province. In particular, the Huzhou center can claim to have a succession of masters from the late Northern Song to the Qing periods, and they also inspired a host of imitators at different times and places. In mirror no. 92, the last in the catalogue, Tang Wanheng, the location of whose workshop is unknown, apparently took pride in his own product and followed in the footsteps of such a great master as Xue Jinhou 薛晉候 of Huzhou.[34]

Ultimately, as tended to be the case during dynastic China, it was the imperial court that, through its long history, continued to make its mark on the mirror. It was supreme political power combined with unbridled wealth that compelled the mirrorsmiths to reach a technical and artistic peak. No other patrons, even such institutions as the Buddhist church and the Daoist temples, could rival His Majesty's ability to inspire and to coerce, albeit benevolently.[35] Signs of such influence were quite apparent. We may point to the superior productions by the Shangfang workshop, which led the way for other private workshops in terms of fine casting as well as design innovation. During the Three Kingdoms period, mirrors were already noted to be a part of diplomacy, as the emperor of the Wei state presented to the Japanese woman ruler Beimihu 卑彌呼 of Xiematai 邪馬台, among other sumptuous gifts, more than a hundred mirrors.[36] During the Tang dynasty, mirrors became Emperor Ming's gift to his favorite consorts and meritorious officials, especially during the so-called *Qianqiu* 千秋 Festival (Festival of a Thousand Autumns), a day to celebrate His Majesty's birthday.[37] Surviving *qianqiu* mirrors can still be found, sharing images with those of non-imperial workmanship.[38] Even more glorious testimonials came from the Shōsōin where, insofar as the Japanese imperial recipient of the Tang's heritage goes, the cherished examples remained unsurpassed in the richness of their repertory, the finesse and sharpness of their images, and their sophisticated and impeccable workmanship and skill.[39] Back in China, another testimonial of a far later date came about when the Qianlong emperor (r. 1736–95) sponsored the remake of the Han mirrors, resulting in a technical tour de force.[40] By that time, of course, the glass mirror had already made its entrance into China, and the venerable tradition of the bronze mirror was not only on the way out from the imperial palaces, but also, broadly speaking, from the land that was the Central Kingdom.[41] As the production of bronze mirrors suffered a permanent setback, they also resuscitated themselves as antique collectibles.

TWO MORE ATTRIBUTES

Two additional attributes of the bronze mirror are worthy of noting. The first is its portability, which can be attributed to its modest dimensions, but which significantly contributes to its mobility. With mobility also comes confusion regarding, in given time periods, the origin of specific works. The second is its value as a commodity, which rendered it a cherished object. Being such, it defied time and may, from the time of its production to its eventual burial, have lasted from one generation to the next. Both attributes pertain to the nature of the mirror itself and affect art historical research.

Portability is an issue in mirror studies because, as a commodity, mirrors can be transported across regions. That the Han mirrors could be found in places as far away as Central Asia, in Siberia, or in Korea and Japan, has been well attested. In the case of the Sino-Japanese link, not only were there diplomatic relations and trades involving mirrors, but there also existed the possibility of traveling mirrorsmiths who made their products in Japan and for Japanese clients.[42] During the Tang dynasty, two identical mirrors made from a single model or mold have been found, respectively, in the tomb of Dugu Sizhen 獨孤思貞 (d. 697, reburied in 698) in Xi'an, as well as in the Takamatsu tomb in Japan.[43] It is also possible to extend our discussion to an earlier time, when the Masters of the Zhang clan (no. 42), likely from the city of Suzhou, Jiangsu province, had their mirrors distributed across an extensive network, which included such provinces as Zhejiang, Hubei, and Hunan as well as, again, Japan. That was a remarkable phenomenon stretching from the period of mid Eastern Han all the way to the Western Jin of AD 273. And they, the Zhang masters, were not the only ones. Others were just as prolific, particularly Chen Shi.[44] A native of Yangzhou, broadly speaking also of the Wu region, he may also have made his way into the Land of the Rising Sun and was a part of the Sino-Japanese exchange.[45]

Beyond the phenomenon of the expanding network in mirror distribution, the main thrust of our concern is the origin of those mirrors during such times as when China was divided, e.g., the Warring States period, or when rival dynasties rose in a Pan-Asiatic context, e.g., the Liao, Song, and Jin. In the former case, with feudal states in contentious struggles, regional styles would seem to have been a critical factor in the history of the mirror. However, mirrors branded as Chu, mainly in the Hunan/Hubei regions, often show up elsewhere—symbolic either of political linkage or a piquant disdain for territorial alignments. The assumption that mirrors are found where mirrors are made, while generally sound, may still require in-depth scrutiny in exceptional cases.[46]

In the Liao-Song-Jin complex of the tenth to thirteenth centuries, the question of origin is even more acute. These were rival dynasties, and their interrelationships were marked by bloody struggles, open warfare, and uneasy peace. The Liao and Song coexisted from the tenth century to the twelfth. Then the Jin overturned the Liao and eventually pushed the Song to the south, resulting in a realignment of power. Still, mirrors from the Song could be easily transported across the boundaries into the Liao and, later, the Jin. Illustrating this northward movement were the famed Huzhou mirrors, whose popularity was both transregional and transnational.[47] Another example was the mirror featuring Ci Fei, a prominent hero in a lost Tang painting (no. 80). In the history of mirror design, this subject emerged during the Northern Song period, and afterwards gained a foothold in popularity north of the Yangzi River. This popularity is evidenced by its being unearthed in the Jin territory as well as in Korea, where it was first imported and then replicated.[48] By contrast, the southward migration of the Liao and Jin mirror types was not so easily documented. Archaeological records indicate that a Jin mirror did make its way into the province of Sichuan.[49]

Given this transregional or transnational character of the mirrors found in excavations, questions do arise pertaining to the origin of *some,* if not all, of these mirrors. In reference to the two northern states, native products existed.[50] We are not contesting their verity. In recent times, for instance, scholars involved in the study of Jin mirrors often thought that the characteristic mirrors of that northern dynasty were those with twin-fish decoration (no. 86) and also those with figures of children (no. 85).[51] If so, they must also explain the fact that the former typology not only had antecedents in Song China, but also could have been produced by Chinese craftsmen for the Chinese population under Jin rule. This would have caused a degree of ambiguity in defining ethnic or cultural characteristics.[52] In the case of mirrors depicting active children, the same set of questions is also pertinent, with an additional query of whether these were also made under the pronounced influence of Southern Song painting or ceramic decoration. In the former context, there is the looming figure of Su Hanchen (active ca. 1101–63), who presided over the raging success of his plump and charming children.[53] In ceramics, one may note that Song porcelains had already incorporated the images of children into their carved design.[54] In the same way, mirrors of floral or avian decoration unearthed at the Liao sites cannot always be thought of as *its* product, for we must also take into consideration whether these could have been Song creations, or Liao replications.[55] This is to say, in order to ascertain the Liao and Jin characteristics in mirrors, we need much more than a geographical fixation of the burials. In the Jin case, the official inscriptions later engraved on the sides of the mirrors can only warrant that these were *used* during that dynasty, but are not sufficient to demonstrate a native production.[56] In short, we are faced with ambiguity regarding the identity of mirrors, such that will demand a thorough and systematic evaluation on a tri-dynastic basis, not the least of which will involve metallurgical studies involving lead isotopes to pinpoint the geographic sources of that component in the alloy of these mirrors.[57]

The second issue lies in the relative "pricey" nature of these mirrors, which assures a longer existence than objects of lesser value. In comparison to ritual vessels, the mirror was of a modest dimension and value. Yet it was not an inexpensive item and, as a commodity, was beyond the affordability of the lower social classes. Even for affluent families, it was used over a lifetime and was not an easily disposable item. As cherished possessions they were sometimes interred with their owners and accompanied them to the netherworld.[58] During other times or circumstances, they could also be turned into family heirlooms and thus transmitted to posterity. *Yi zisun* 宜子孫 (Suitable for your sons and grandsons) was a phrase that appeared in mirror inscriptions with a degree of frequency (e.g., no. 17).

Regardless of the twofold possibilities—either being buried with the owner or serving as a family heirloom—we may safely assume that a possible discrepancy exists between the time of a mirror's casting and the time of its final disposal. If it were buried with its first owner, it would have been preceded by a normal period of usage, say, from less than a decade to possibly two or more. Just as in the case of ritual bronzes, the dates of their burial could not be but a very rough *terminus ante quem* for their casting.[59] In the case of an heirloom, a delay in final disposition would extend the interval further and introduce cross-generational issues. We are, of course, discounting those mirrors that were made specifically for burial: those that are little more than a thin disc, cast in a perfunctory manner, and which survive in large quantities.[60] Hypothetical as all this may sound, the key lies in our ability to document actual cases where delayed burials of the mirrors are more than mere speculation.

A clear example of where a time frame for the creation and disposition of a mirror can be established is the mirror dated by inscription to 1591 and found in the tomb of Wang Xijue (1534–1610) and his wife (d. 1598).[61] If the mirror had belonged to the famous Ming official/scholar himself, then a span of nineteen years intervened between its production and final disposition. If the mirror was Madame Wang's, then a shorter interval of seven years would prevail.

On the other hand, if this example is considered as being too late to be typical, then we can cite the following instances for belated entombment of mirrors. A Tang animal-and-grape mirror was found in Liu Zizhen's burial, dated 851.[62] This find was anachronistic, inasmuch as the main cluster of this mirror type has been dated to a time period roughly between 664 and 711. Hence, a delay of no less than 140 years has taken place at its final disposition. Another Tang mirror, similar to no. 62 in the catalogue, has been found to accompany the burial of Yan Wan, dated to 725.[63] In view of the simple image in the mirror itself, versus the far more elaborate mirrors that were being made during the second half of the seventh century as well as the first quarter of the eighth, the date of the burial actually implies a similar anachronism. Having married into the imperial clan, with ample financial resources to procure any mirror she desired, there was no apparent reason for her to keep this simple, archaic instrument unless it had held special, personal value. By all indications, this particular mirror should carry a date more than a few decades earlier (perhaps ca. 620s)—back to the time of Yan Lide, her father, who may have designed it.[64]

On a transdynastic note, the continual appearance of Tang mirrors in later tombs has been documented. An instance of that occurred in a Northern Song tomb; the mirror was one of the lobed varieties with floral and avian motifs.[65] Among Liao burials of eminent personages, we may mention at least three notable examples where early mirrors, again of the Tang pedigree, made their presence known: a Liao official's tomb at Yemoutai, the 942 tomb of Yelu Yuzhi (890–941), and the combined burial of Princess Aoge and her husband, dated 959.[66] The Jin also displayed a similar reverence toward the early heritage; as an example, one burial site (dated 1199) contained an animal-and-grape mirror of Tang vintage.[67] In these and other instances, the long stretch of time may have turned some of the mirrors into "antiques," to be collected and cherished. A case in point pertains to yet another animal-and-grape example found amid a hoard of Liao silver objects in Inner Mongolia, testifying eloquently to the added value beyond its functionality.[68] If this single instance does not attract our attention, then the hundred or so ancient mirrors, mostly of Han and Tang vintage, in the Song imperial collection should make us aware of the scale of the collecting and the passion and dedication involved—the like of which had not been attested to prior to this time.[69] One may assume that a majority of the mirrors published in the imperial catalogue of *Bogutu* 博古圖 would have been rescued from underground.

Given our awareness of delayed burials, we must therefore approach archaeological finds with a degree of caution. Even when facing sites with numerous burials, where a relative chronology can be devised (e.g., Shaogou near Luoyang and Guangzhou in the south), one should still be circumspect regarding whether the same chronology can always be effective with respect to the mirrors interred therein. In the same way, unless it is fully justifiable, it may be prudent to avoid either dating the tomb by the mirror(s) found within, or dating the mirror in association with objects within the same burial chamber. The assumption of contemporaneity may be enormously tempting, but one can only do so at one's own risk.

By the same token, were we to approach the art of the mirror by type, it is possible to find a few signs and clues of the delayed burial syndrome. A case in point is the TLV mirror. As far as this typology is concerned, there is little question that by the end of the Western Han period most, if not all, of the subvarieties of the TLV mirror had reached maturation.[70] If so, then the continuance of these very same types into the Eastern Han and beyond—for a duration of approximately two hundred years—cannot always be explained away on the grounds of continual popularity and, therefore, sustained efforts at replication. We may point to the presence of the octet arrangement in the TLV mirror in Eastern Han burials in Guangzhou, which elsewhere could boast a time frame of Xin period or earlier.[71] Another case pertains to a type of TLV mirror with nothing more than curlicues (e.g., no. 25); in an excavation report on Guangzhou sites, these mirrors are listed under both the early or later phases of Eastern Han,[72] whereas something similar unearthed in Luoyang has been dated to the Western Han period.[73] Granted, the farther the mirrors were from the center of the empire, the greater the chronological gap would be. In view of the general contention that in Guangzhou mirrors were imported, this could conceivably increase the range of factors that would contribute to the delay in their arrival in this remote city and then, in turn, for the postponement in their final disposition.

What is the general impact of this belated burial syndrome? For one, it serves to distort the chronological sequence of the archaeological finds. Granted, the distortion may be less than severe; in time, collaborating evidence can perhaps lend a hand in straightening out a temporal profile that may have been compromised. But it does highlight the need for caution, especially when the art historian has no other recourse than to date the mirror by the site of entombment. In the case of TLV mirrors, some redating may be necessary, even critical. This pertains to those mirrors with the inscription that begins with the familiar assertion 漢有善銅出丹陽 (During the Han period, fine copper is being mined in Danyang).[74] The question is, does the "Han" here refer to Western Han or Eastern Han? If we are fairly certain to lay claim to a

Western Han provenance for production when such mirrors have been unearthed in pre-Xin and Xin burials, can we be equally confident in providing an Eastern Han period date for similar mirrors from the burials within the latter time frame? Here, the belated burial syndrome renders uncertain what was once deemed as definitive.

In a broader sense, this means that the logic of the evolution in mirror types is not always in tune with the chronological framework derived from archaeology. In their divergence, the main culprit consists of the accidental nature of the finds, which is beyond our control. A lesser culprit is the belated burial syndrome, which creates further distortion, but at least it can be partially remedied by means of stylistic and iconographic correctives. Only awareness of the extent of its impact will allow us to deal with it.

On page 23 begins the catalogue proper, with 92 entries divided into five broad phases. Phase I comprises nine mirrors of the Warring States period (475–221 BC). Phase II spans the period from the Han to the Northern and Southern dynasties (206 BC–AD 589), featuring thirty-six mirrors. Phase III includes mirrors of the Sui-Tang eras (AD 581–907), with thirty-one examples. Phase IV encompasses the triad of the Liao, Song, and Jin dynastic houses (916–1279), and the mirror production in all three includes eleven examples. The last phase, or Phase V, pertains to the length of time which we shall summarily call post-Song (1279–1911), with a limited sampling of only six mirrors. This phase-by-phase account is devised here since the objective of this catalogue is geared not to an academic survey of the mirror's evolution, but to an overview of a collection, admirably formed through a judicious balance of historical perspective, aesthetic sensibility, and keen connoisseurship.

J. C.

1. Prior to the Carters' donation, CMA's records showed nineteen mirrors altogether. Of the nineteen, ten were acquired in 1915, two in 1926, and the rest spread between 1972 and 1995. I am grateful to Nancy Grossman for providing this information.

2. For a complete list of the Carter collection, see the Appendix in this catalogue.

3. See the preface to Rupert/Todd 1935, where O. J. Todd states that this was "based on a collection of more than one thousand Chinese mirrors. . . ." The catalogue entries contain a total of 429 mirrors. Comparable to the Carter collection is the Donald H. Graham Jr. collection now housed at the Honolulu Academy of Art; see Nakano et al. 1994, which contains 117 mirrors in its catalogue entries.

4. See He T 1990, 74–79, especially p. 78.

5. It is usually assumed that the mirror with handle became known in China during the Tang dynasty, and was popular during the Song period and afterward. Nie Shimei (Nie 1994, 65–66) argues that such mirrors were used during the Han dynasty. His sources, however, are questionable.

6. Here we may add a note that, in some literary sources, it was reported that some bronze mirrors might have reached dimensions as large as 5 x 4 Chinese feet, much like today's glass mirrors; see Nie 1994, 65–68.

7. For a simple illustration, see the Wu family shrines, long noted for their thematic richness; see Wu H 1989, fig. 41, for the divine pair. For Fuxi and Nuwa, it may be useful to check the compilation in Wen 1956, pl. 28. For Four Spirits, ibid., pls. 50 (Blue Dragon), 19–20 (Red Birds), 51 (White Tiger), and 49 (Dark Warrior).

8. See the catalogue entry for mirror no. 80.

9. See the catalogue entry for mirror no. 62.

10. MacKenzie 1999, 82–91.

11. For *liubo* game board, see Xiong 1979, 35–39; Fu 1986, 21–33; Zhou Z 1987, 1116–18; Cheng Z 1997, 124–26. For the simplest TLV mirror dated to Western Han, see Ma et al. 1998, pl. 57, unearthed from Changsha, Hunan.

12. See Cammann 1948, 159–68.

13. See Tseng Yu-ho Ecke's essay in Nakano et al. 1994, 46–49.

14. See Cheng Z 1997, 124. The mirror bearing the inscription came from a Western Han burial. For a different version, see Kong 1992, 266. The mirror is dated to the Xin period; also reproduced in Zhou Z 1987, fig. 1.

15. Speaking in metallurgical terms, the two broad phases, insofar as bronze alloy composition is concerned, are (1) the high-tin, low-lead formulae, covering the Warring States to the Five Dynasties and also a part of the Song period, and (2) the low-tin, high-lead formulae, covering the Song and post-Song periods. For percentage figures, see He T 1992, 33–48.

16. See the argument presented in O'Donoghue 1990, ch. 4; for horse and chariot fittings, see ibid., figs. 27, 29, 32, 34, 36, 37, 39, 40.

17. For religious (primarily Daoist) messages, see Cahill 1986, 62–70. In terms of ancient scripts, Zhou SR 1986a is unparalleled for its analytical acumen.

18. For Shandong example, see the relief in Wu family shrine in Omura 1915, vol. II, pl. 164 (middle section). For relief in Sichuan, see Wen 1956, pl. 55.

19. We are referring to the primary function of the mirror: to reflect the image of its owner or those who should come to view their own visages in it, either narcissistically or as a corrective device to modify one's appearance. Other functions of mirrors—e.g., as a token of affection and love, in burials, on the containing head mask, and lastly, as architectural decorations, etc.—are but an extension of that primary function. For mirrors used as symbols of affection, especially in the case of mirrors broken into two halves, an excavated example has been reported by Peng/Tang 1980, 28ff, especially pp. 28–29, fig. 16, where the husband and wife were buried each with a half of the mirror. The use of mirrors attached to the masking box that enclosed the deceased's head began as early as the Western Han period; see, for example, Huaiyin 1988, 59–67, especially pp. 63–64, fig. 8. For architectural use of mirrors, see the example of the White Pagoda, Chongshou Temple, in Shenyang; Li C 1957, 34–38. Here we touch only tangentially on the mythic function of the mirror, as an instrument to unveil the inner organs or to unveil the true nature of man or other creatures who, in tales and legends of the past, assumed the appearance of human beings.

20. There are indications that male and female users of the mirrors may favor different types. In the tombs of Liu Sheng (d. 113 BC) and his wife, Dou Wan (d. 104 BC), the former, the Prince Jing of Zhongshan, was accompanied in death by a type of mirror similar to no. 11 in this catalogue, while his spouse was buried with an early TLV mirror (cf. mirror no. 24) and also a larger and far more richly decorated example than mirror no. 12 in this catalogue; see the catalogue entries for these mirrors. In a report on the Jin burials at Nanchang, Jiangxi province, it is recorded that the husband used a mirror with four leaves and phoenixes and birds design, whereas his wife favored a mirror with rings of squares and semicircles, its inner compartment peopled with deities and animals; see Kiangsi 1974, 373ff, fig. 3 on p. 374 for the wife's mirror, and pl. 10-(2) for the husband's.

21. There is, of course, the depiction of the suicide of Wu Zixu, the loyal official in the court of Wu during the Eastern Zhou period; see Chen P 1987, pl. 51, and explanatory text, p. 30. This is not so much a celebration of violence as a comment on the folly of a ruler who refused to heed sound advice and condemned a wise counselor to death. The result, as predicted by Wu Zixu, was the eventual downfall of the state of Wu. For a new perspective on this issue, see Wang E 1994, 511–34.

22. For early history of Chinese bronze mirrors from the Qijia culture onward, see Kong/Liu 1994, 8–18; see also O'Donoghue 1990, 5–183. At this time, I do not wish to become enmeshed in discussion of the origin of the mirror in China, whether it is indigenous or imported.

23. In this catalogue, no. 37 is given the name of *Mirror with Quatrefoil, Paired Phoenixes, and Auspicious Animals*.

24. The question of typology remains unsettled and will require a careful scrutiny over established conventions. Today, criteria for the known typology are varied and marred by discord. Sometimes the inscription becomes the basis for establishing types, e.g., the *riguang jing* versus *zhaoming jing*, whereas the design of these types is essentially identical. Other times the design of the images contributes to the typological distinction, e.g., TLV mirror, without referring to the inscription data. Concern must also be voiced that, once the types

are cast, we should not overrate their distinctiveness but instead also observe their linkages and their indebtedness or impact, whether mutual or partial.

25. For examples of this type, see Chen P 1987, pl. 56. Here there is represented, from high to low, a full range of Daoist deities.

26. Interestingly, the square inside a circle modality did not vanish, but resurrected itself time and time again. See, for instance, Song and Liao reinvention of the pattern, with or without cosmological implications. For Song examples, see Kong 1992, 762–65. For Liao specimens, see Liu S 1997, pls. 27, 28, 30, 31, 40–42, 45, 47–53, etc.

27. For this inscription, see Kong/Liu 1994, 103.

28. The lobed mirror was invented no later than 670; see Yen 1989, 289–366, especially p. 296, relative to the bronze mirror found in Shaanxi province (Shaanxi 1959, pl. 150). In the case of those with foliated contour, the earliest appearance can be dated to AD 707; see Kong 1980, 396.

29. Yuen 1986, 145–61; for the above-mentioned shapes, see illustrations on pp. 152, 154, 156, 157.

30. See the translation of the Shangfang provided in Hucker 1985, 407: "Directorate for Imperial Manufactories."

31. Chang'an and Luoyang were of course capitals, respectively, of Western and Eastern Han. Excavated works from these sites are plentiful; see Shaanxi 1959 and Luoyang 1986. Echeng was made the capital of the Wu state during the Three Kingdoms period; see Hubei/Echeng 1986, 1–9 ("Introduction" by Yu Weichao 俞偉超), and 10–16 (general essay by Guan Weiliang 管維良).

32. For Suzhou and Shaoxing, see Wang ZS 1985, 1025–31.

33. See Bian 1977, 31–37; Zhou/Zhou 1979, 53–58.

34. See the catalogue entry for mirror no. 92.

35. For the Buddhist church, mirrors have also been placed inside sculptural portraits of arhats; see Jinan et al. 1984, 76–82. In addition, Buddhist monks and Daoist priests utilized mirrors for a number of ritualistic and religious purposes.

36. See Pei 1967, vol. 6, 4629–32, especially p. 4631, where it was mentioned that the Wei court twice sent mirrors to Beimihu, among other things. The first time was in AD 238 (or AD 239), involving hundreds of mirrors, and the second time in AD 241. For the controversy surrounding this matter, see Wang ZS 1981a, 346–58.

37. For example, see Emperor Ming's poetic testimony, included in Zhonghua 1960, vol. I, 32, where His Majesty bestowed mirrors upon his consorts and also his officials during the festival called *Qianqiu* (Festival of a Thousand Autumns). Ibid., vol. I, p. 41, refers to perhaps a Festival of a Thousand Autumns in a different year when mirrors were distributed among officials as gifts. Also ibid, vol. I, p. 33, the emperor received a gift of a sword and mirror from a Daoist priest, Sima Chengzhen.

38. See Taipei 1986, pl. 114.

39. For examples, see Shōsōin 1994–97, vol. I, 74, 76–79, 82; vol. VII, 84–88.

40. See Guo 1996, pls. 186–97.

41. See the glass mirror reproduced in Taipei 1986, pl. 116. This has a cloisonné backing.

42. Regarding this issue, Wang Zhongshu 王仲殊 was, among Chinese scholars, the strongest advocate for traveling mirrorsmiths from Wu China to Japan; see Wang ZS 1981a, 346–58; and Wang ZS 1984a, 468–79.

43. See Wang ZS 1981b, 277–78, figs. 1, 2; and Wang ZS 1982, 410–13.

44. See Wang ZS 1986, 1017–25.

45. See Wang ZS 1984b, 1118–26.

46. See Liu Y 1985, 1008–14.

47. This is best attested in Huzhou mirrors found in North China; see Zhang Y 1990, pls. 22–24.

48. See the catalogue entry for mirror no. 80.

49. See Cheng TK 1948, 86–88, especially p. 88, recounting how a Cheng'an era mirror of 1198 was found near Chengdu, Sichuan. The tomb also yielded a real estate transaction dated to 1212.

50. For impeccable Liao mirrors, see Liu S 1997, figs. 1, 182; also figs. 183, 184. The former set is dated to the Liao period, and the latter set bears inscription in Qidan script. For Jin mirrors, there are also dated examples; see Hebei 1996, pls. 228, 244, 256, 262, 305, 319, etc. These are either dated or bear the coinage of the Jin dynasty. Regarding Jin mirrors, one must be cautious with respect to those with marks of official sanctions that were engraved on the rim, as these marks indicate official approval of usage and sale, but do not indicate origin.

51. See Wang/Li 1987, 365–80, especially pp. 371–73.

52. Twin-fish design can be traced back to the Han period; see a bronze washing basin excavated from Santai Xian, Sichuan, published in Santai 1976, 395. During the Song period, a number of datable mirrors also bear this design pattern; see Zhang Q 1992, 21–24. On p. 23, the author mentions a twin-fish mirror from the tomb of Chen Anzu 陳安祖, who was buried in the fourth year of Congning (1105); the mirror is reproduced in fig. 10 in the back of the cover. See also Wang SL 1987, fig. 166, for a Huzhou twin-fish mirror made in the Xu family workshop. Also of interest is a mirror with a handle found in a Northern Song burial, which displayed the impending metamorphosis of the carp into dragons; see Luoyang 1992, 37–51, figs. 27-(5), 37.

53. See National Palace 1989, 67–68. The painting is entitled *Children at Play in an Autumnal Courtyard.*

54. See Wirgin 1970, figs. 16–19. Also, more directly, the Jin mirrorsmith may have been indebted to a Tang counterpart; see Carter mirror CMA 1995.343 in this catalogue's Appendix.

55. For floral motifs, see Liu S 1997, figs. 79–126. For avian motifs, see for instance the rendering of phoenixes in ibid., figs. 130–35.

56. See Kong/Li 1992, 170–74.

57. See Peng et al. 1985, 1032–37.

58. The majority of the burials contain one or a limited number of mirrors. In the Shaogou sites, Luoyang, one particular burial, No. 95, contains 127 mirrors; see IOA 1959, 160.

59. Han mirrors have been found in Tang tombs just as Tang mirrors have been found in Liao and Jin tombs; see Wu W 1991, 188–90, fig. 3.

60. An example is the Carter mirror CMA 1995.316; see this catalogue's Appendix. A brief survey through archaeological reports could show that this type of simple disc extended from the Eastern Han to the Six Dynasties.

61. See Suzhou 1975, 51–56 (not illustrated).

62. See Yen 1989, 335, chart 6, no. 16.

63. See Hubei/Yunxian 1987, 30–42.

64. See the catalogue entry for mirror no. 62; see also Chen GH 1987, 35–39.

65. See Haining 1983, 31, fig. 16.

66. For these three sites, see respectively the following: Liaoning 1975, 26–36, fig. 8; Mongolia et al. 1996, 4–32; and, for Prince Aoge and her husband's tomb, a brief summary in Liu S 1997, 8–10, fig. 5.

67. See Tengxian 1984, 349–51, figs. 2, 6b.

68. See Ba'youwen/Chengshun 1980, 45–51, fig. 16-(4).

69. See Wang F et al. 1969, *juan* 28, 2159ff.

70. There are several types of TLV mirrors: (1) Plain TLV mirror, a simple adoption of the *liubo* board, with no additional decoration, datable to the Western Han; see Ma et al. 1998, pl. 57. (2) TLV with *panchi* pattern as in mirror no. 24 in this catalogue, with a date of no later than mid Western Han. (3) TLV with Four Spirits inserted into the horizontal bars of T's and L's, datable to the Xin era but conceivably far earlier; see Kong 1992, 296. Another intriguing specimen is in the collection of the Palace Museum, dated to AD 136; see Guo 1996, pl. 62. (4) The classic eight-figure arrangement, with the Four Spirits accompanied by four additional figures. A number of these are self-dated to Western Han or the Xin periods; see Luoyang 1996b, fig. 27, dated to 15 BC; Chen P 1987, pl. 38, dated AD 15; Ma et al. 1998, vol. 16, pl. 59, dated AD 10; ibid., pl. 60, dated 15 AD; Kong 1992, 266–67, both dated to the period of Xin. Other publications also feature Western Han specimens; see Jingzhou 1995, 985–96, fig. 9. The last known example with a clear chronology is the mirror dated to AD 235, or the third year of Qinglong, in addition to a few other examples found in the Western Jin burials or later; see Wang ZS 1994, 727–35. (5) TLV mirror with eight-bird design; see Kong 1992, 301, said to be from a Xin burial. Another specimen from an early Eastern Han burial can be found in Guangzhou 1981, vol. II, pl. CXI-2. An even later work, of Southern dynasties, appears in Hengyang 1986, 1079–93, fig. 19. (6) TLV mirror with revolving clouds around nipples and also with curlicues. For Western Han specimens, see Luoyang 1986, pls. 22, 23. (7) TLV mirror with geometric pattern; see ibid., pl. 25, dated to Western Han.

71. See Guangzhou 1981, vol. I, figs. 276-(2)–(4), all of which are later Eastern Han vintage; for early Eastern Han examples, see ibid., figs. 209, 210.

72. Ibid., vol. I, fig. 210-(4); vol II, pl. 169-(6).

73. See Luoyang 1986, pl. 24.

74. For examples of mirrors with "In the Han dynasty, good copper is being mined at Danyang," see Zhou SR 1986a, figs. 58–61. All these mirrors appear to have been dated by burial sites to the early Eastern Han dynasty. By our argument, some of these mirrors could have been works of late Western Han. Incidentally, a parallel construction can be seen in this catalogue's mirror no. 27, where the inscription begins with "In the Xin dynasty, good copper is being mined at Danyang." The latter inscription poses very little problem since the period of Xin is short, whereas the Han period is long.

Plate 1

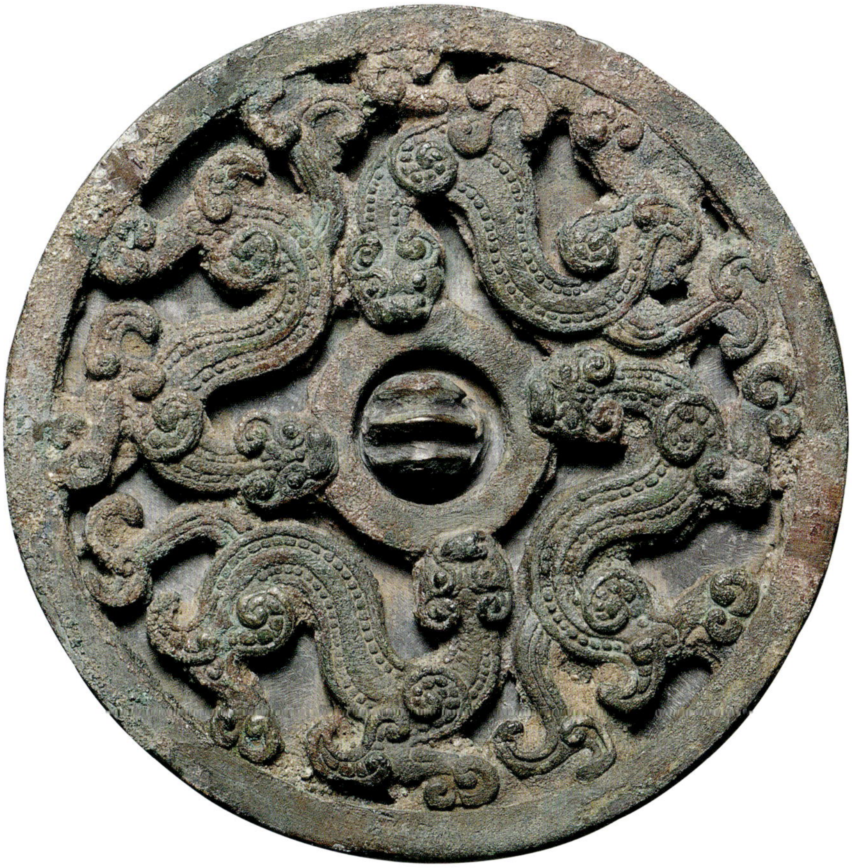

透雕紋鏡
Double-tier Mirror
Warring States (475–221 BC)
Cat. no. 4
1999.220

Plate 2

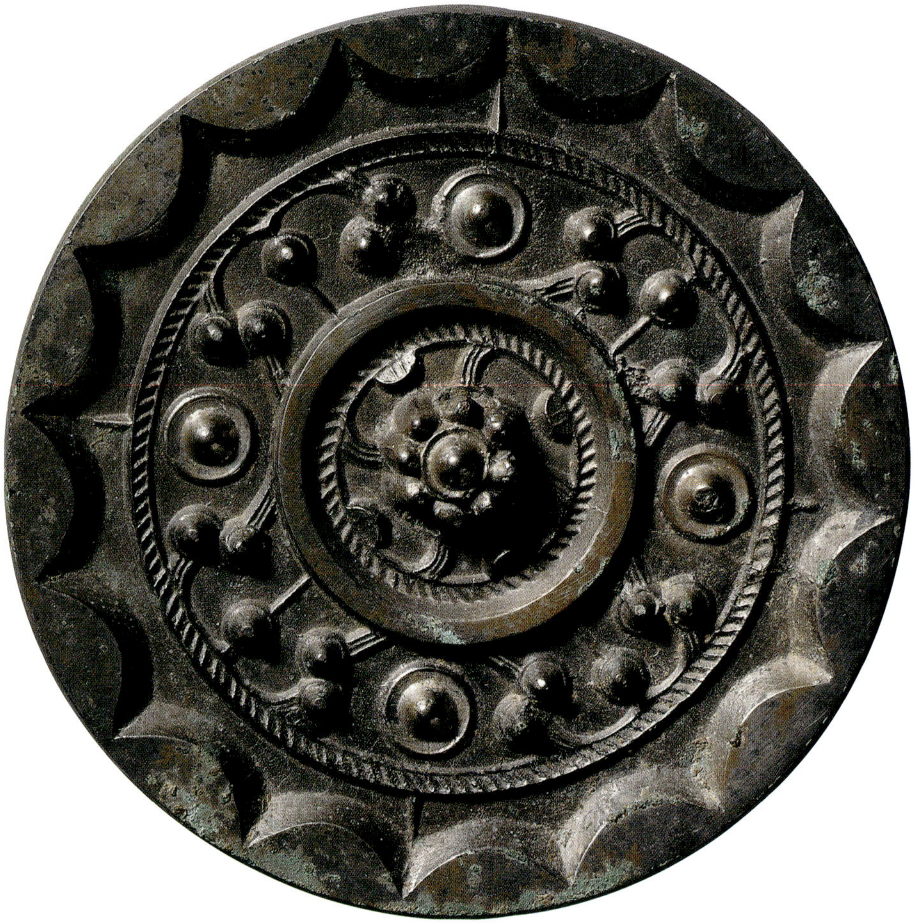

星雲鏡
Mirror with Clouds and Nebulae
Western Han (206 BC–AD 8)
Cat. no. 13
1995.288

Plate 3

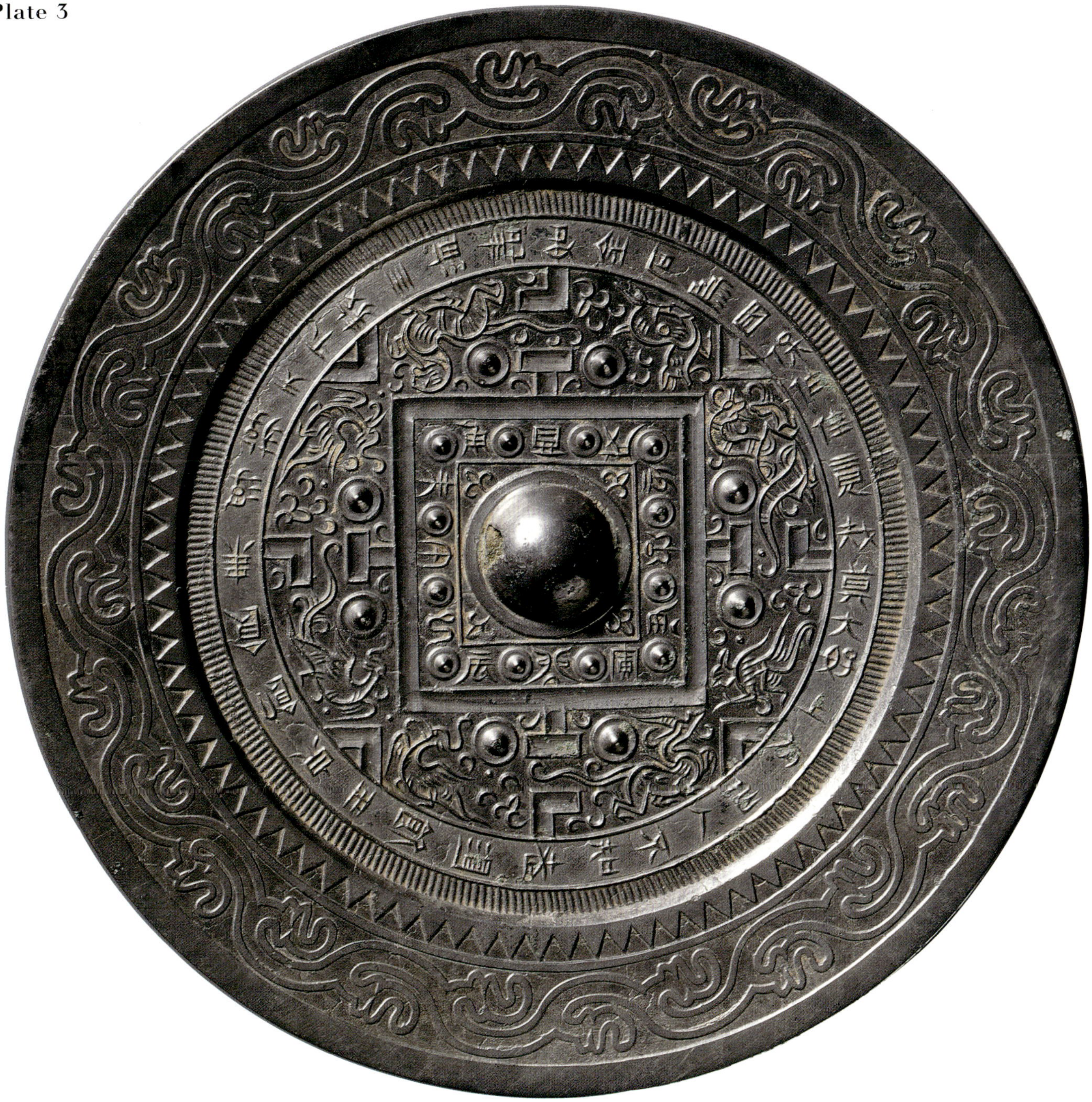

多乳博局式鏡
TLV Mirror with Multiple Nipples
Xin (9–24)
Cat. no. 28
1995.298

Plate 4

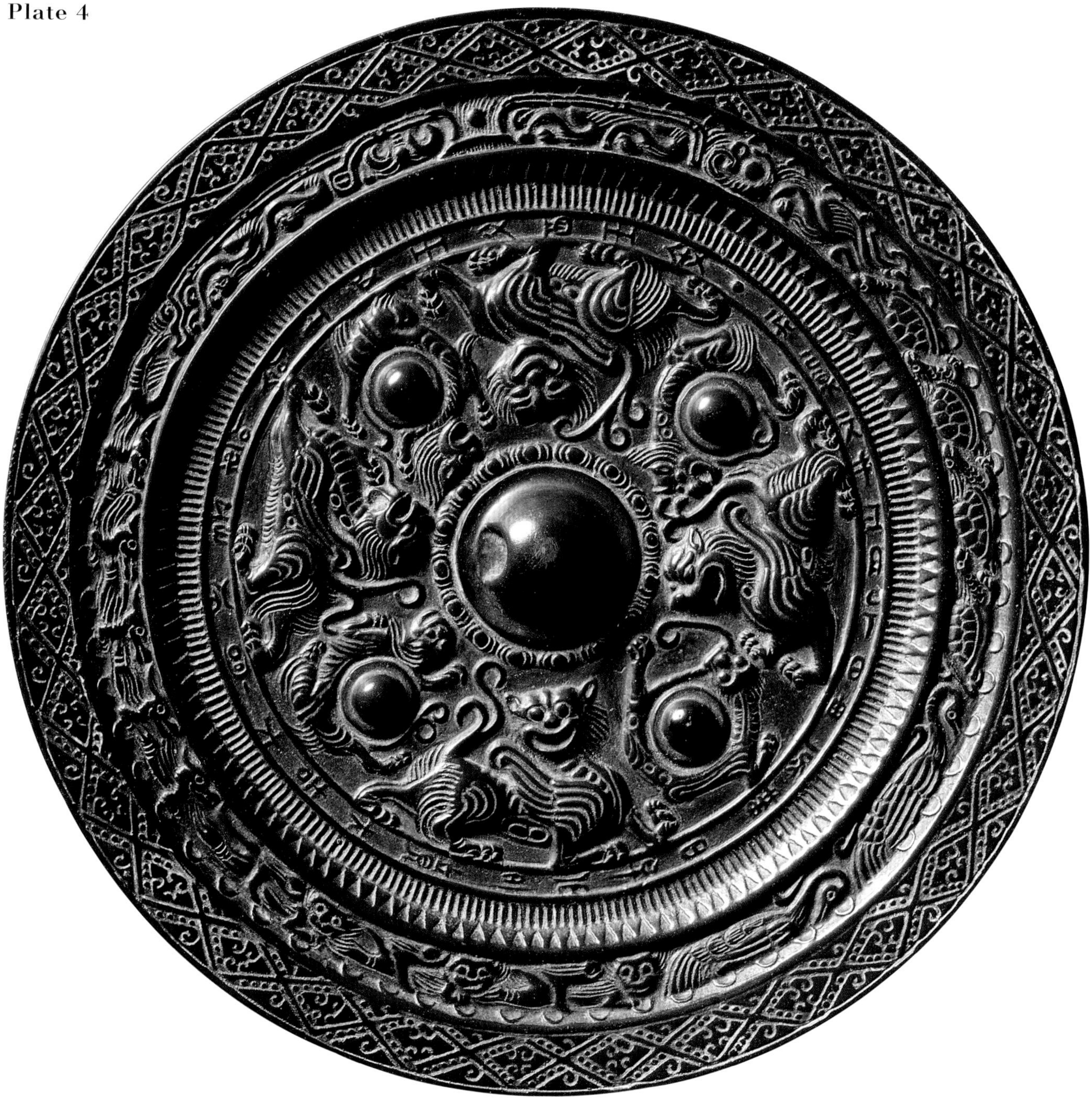

四乳八獸鏡
Mirror with Four Nipples and Eight Animals
Western Jin (265–316), dated AD 300
Cat. no. 44
1995.334

Plate 5

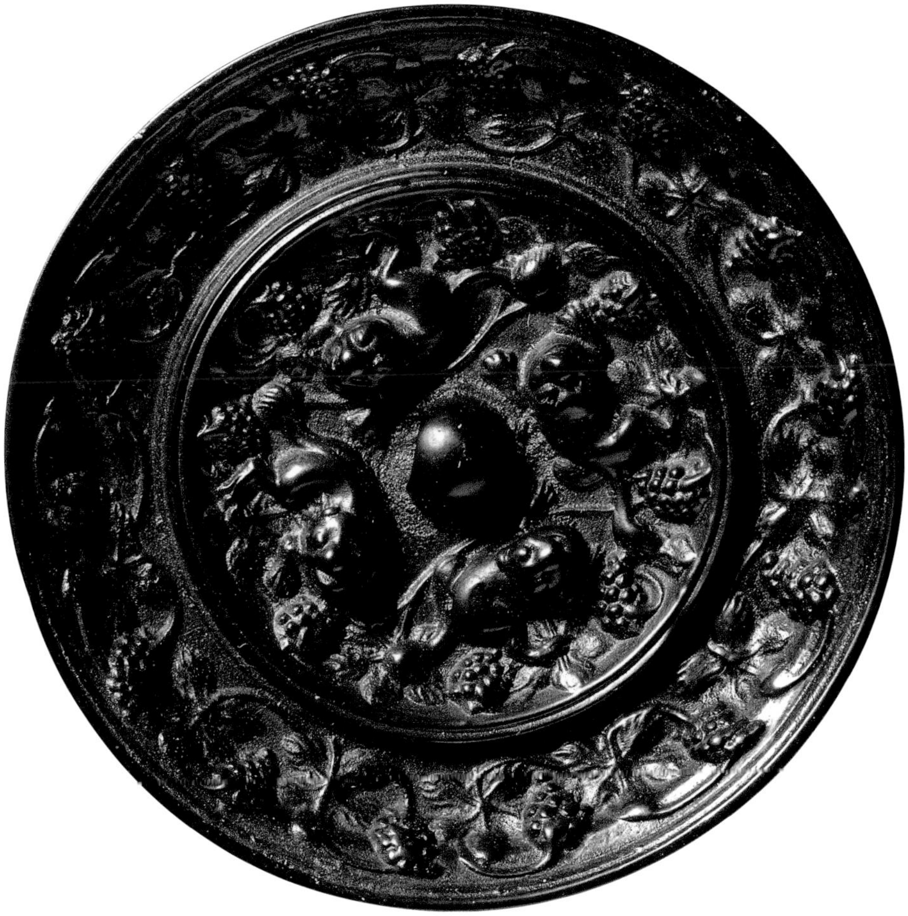

瑞獸葡萄鏡
Animal-and-Grape Mirror
Tang (618–907)
Cat. no. 53
1995.356

Plate 6

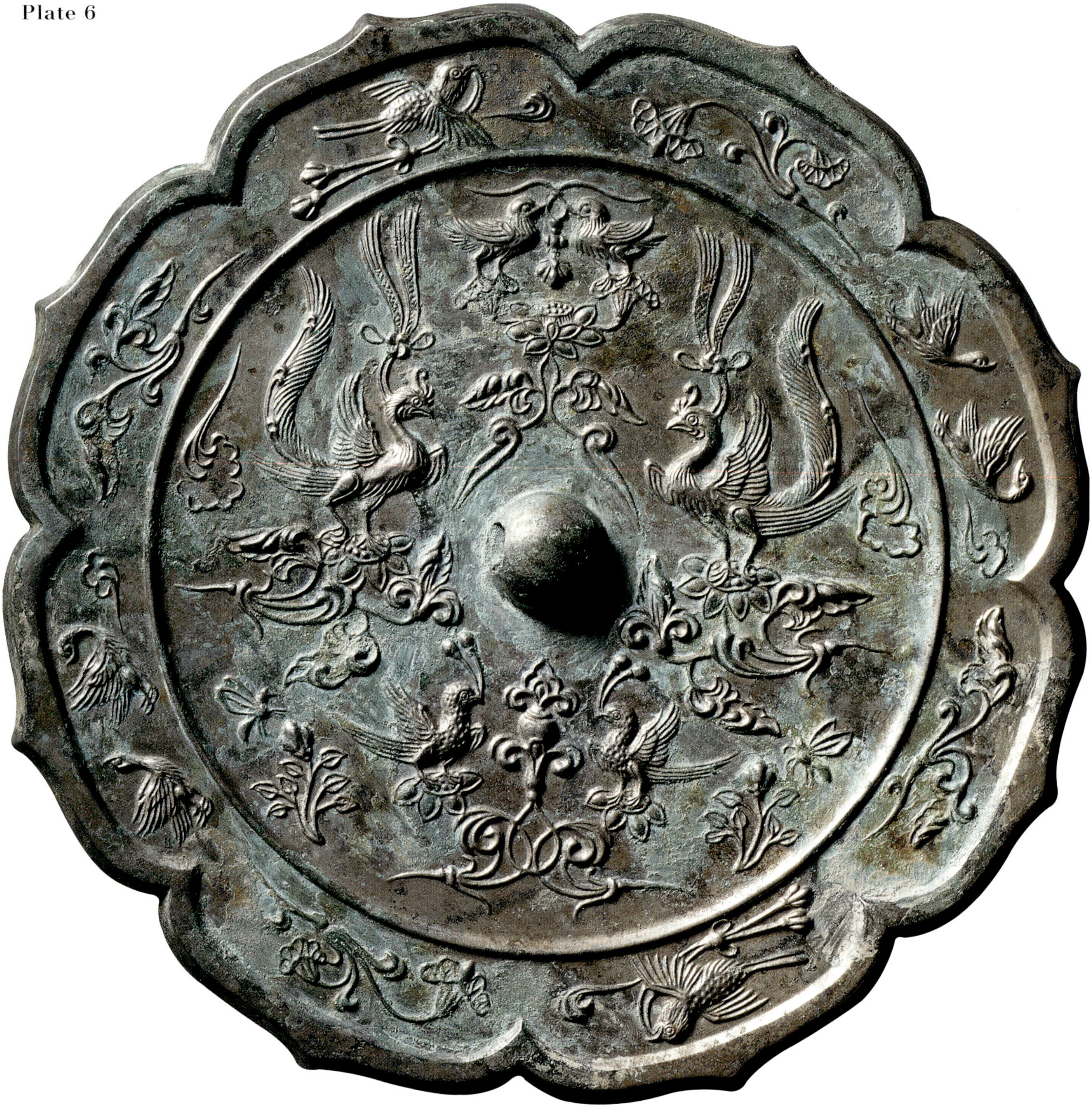

雙鳳花鳥菱花鏡
Octafoil Mirror with Paired Phoenixes, Birds, and Flowers
Tang (618–907)
Cat. no. 68
1995.366

Plate 7

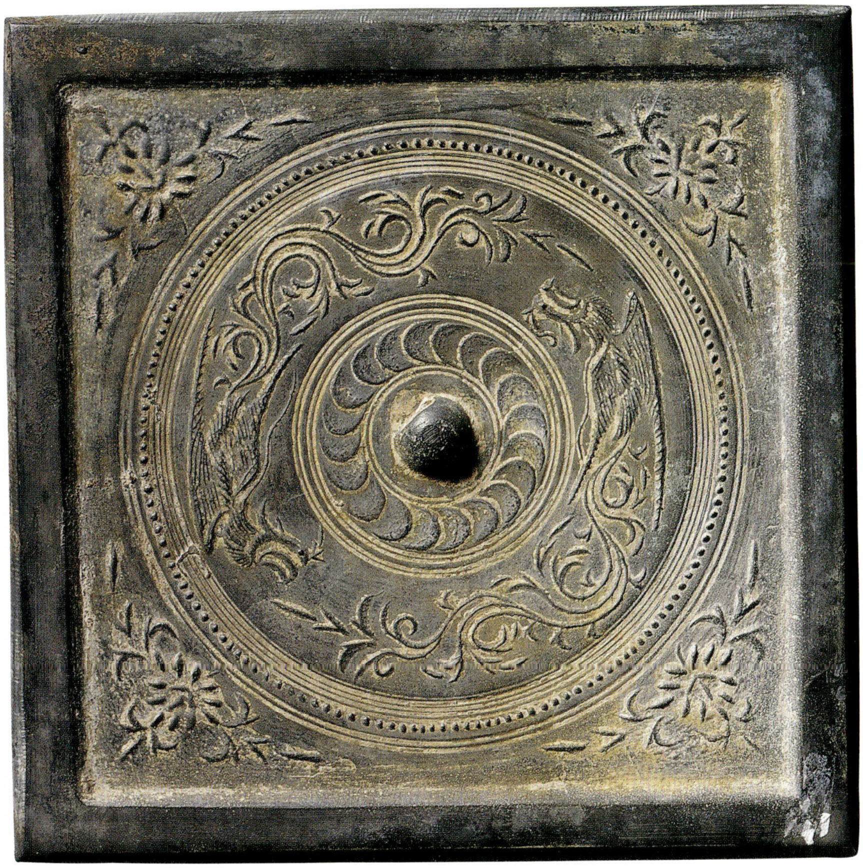

雙鳳花枝方鏡
Square Mirror with Two Phoenixes and Floral Sprays
Northern Song (960–1127)
Cat. no. 77
1995.374

Plate 8

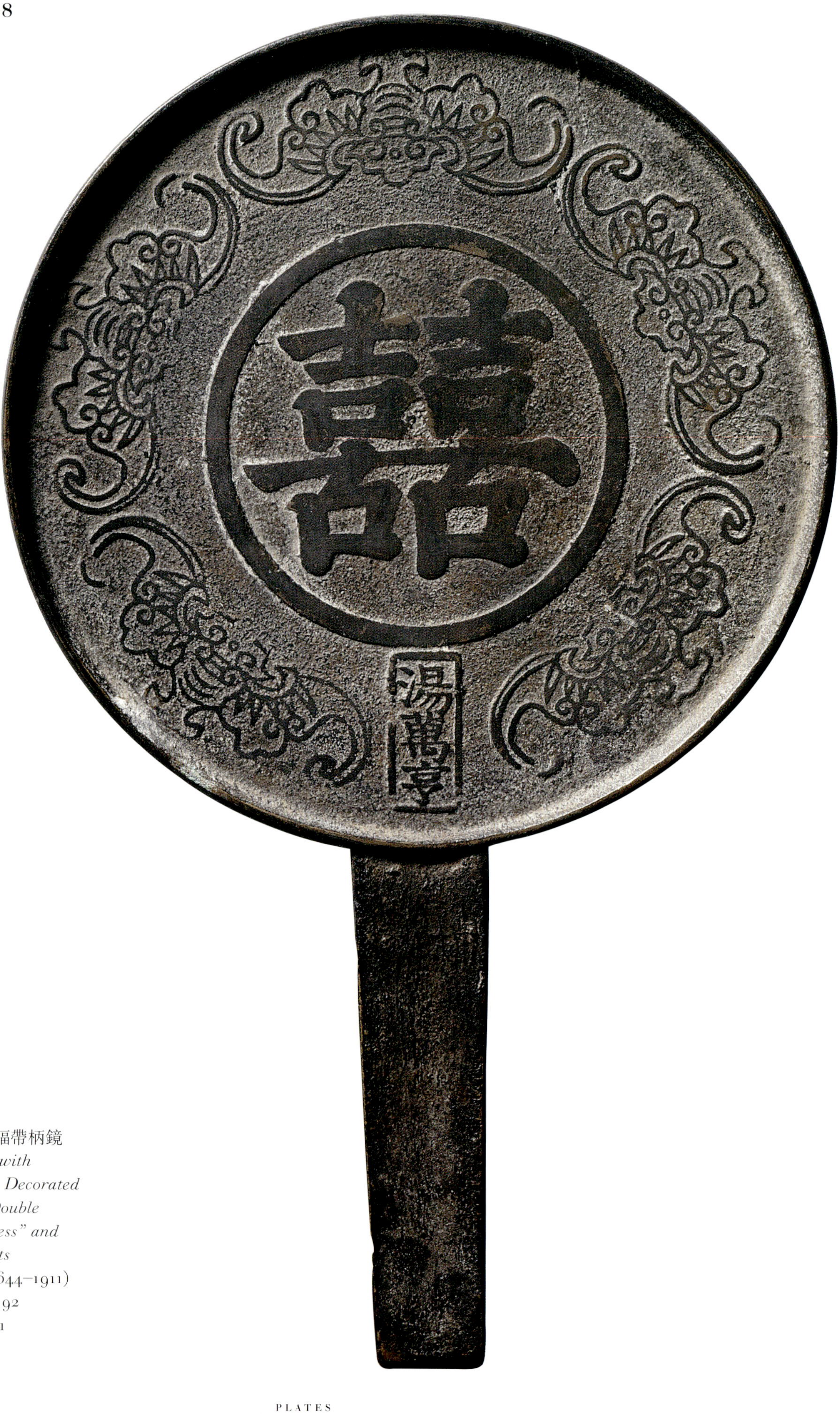

雙喜五福帶柄鏡
Mirror with Handle, Decorated with "Double Happiness" and Five Bats
Qing (1644–1911)
Cat. no. 92
1995.401

Phase I

Warring States Period

475–221 BC

1.

四葉羽紋鏡

Mirror with Quatrefoil and Feather Pattern

Warring States (475–221 BC), 4th century BC
Weight: 90.5 g
Diameter: 11.2 cm
Rim height: 0.35 cm
Overall height (knob): 0.6 cm
1995.267

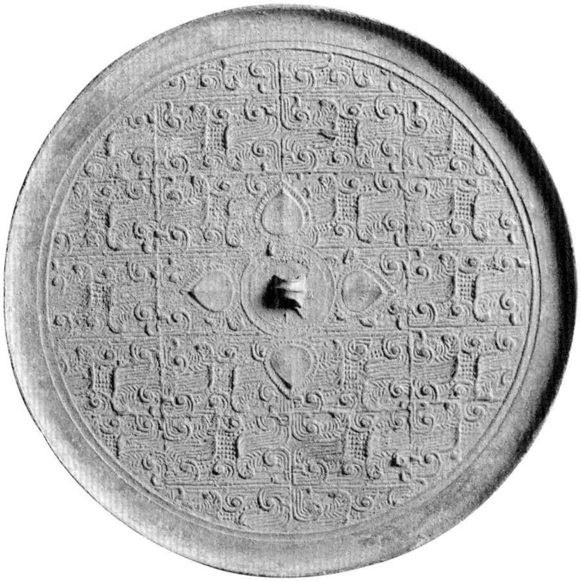

This round mirror has a beveled edge. Its plain border encloses a field of rectangles which is decorated with a feather-hook design surrounded by swirls and granulation. Each rectangle is a modular unit, identical with the rest. During the production process, the mirrorsmith clearly used a stamp to impress the design a number of times on the mold or model. This field of stamped designs then constitutes the ground pattern, the spread of which is limited only by the mirror's allotted surface. Above this ground pattern, around the small triple-fluted central knob, is a circular seating from which the quatrefoil's double petals radiate in four axial directions.

The dual structure of an overall surface pattern and the superimposition of a set of motifs atop is basic to the design of mirrors during this period. Very possibly a product of the Chu region, the Carter mirror is light in weight and in appearance, with a delicate tone that permeates the entire design.

For comparative purposes, we may point to an identical mirror found in Echeng, Hubei province.[1] Another comparable example was unearthed in a mid Warring States site in Hunan province.[2] A site datable to the mid Warring States period draws only a lower limit for the tomb content within; its own dating does not exclude the possibility that the mirror may have been made earlier.[3] After all, permutations of the same design, such as the enlargement of petals in mirrors that represent a later phase of this typology, have also been found in the mid Warring States sites.[4] The question is then, how much earlier can we date the origin of this typology, or that of the individual piece? A difference of two to three decades still may not break through the temporal confine of that "middle phase." Other permutations, consisting of changing shapes or increasing complexity, further explore the potential that is unveiled in the typology.[5]

1. Kong 1992, 23.
2. Ibid., 24.
3. See Liu Y 1985, 1009. Liu believes that such a dual-structure design appeared in the Chu region as early as the Chunqiu era, whereas in other regions its emergence was delayed until the early Warring States period and only became popular still later. See also Lei 1982, 20–36, especially the chart on p. 28, devising a chronological span for each of the mirror types for the so-called Chu mirrors.
4. Kong 1992, 25.
5. Ibid., 26–36.

2.

四山鏡

Mirror with Four T's

Warring States (475–221 BC), 4th century BC
Weight: 124.5 g
Diameter: 10.4 cm
Rim height: 0.45 cm
Overall height (rim): 0.45 cm
1995.281

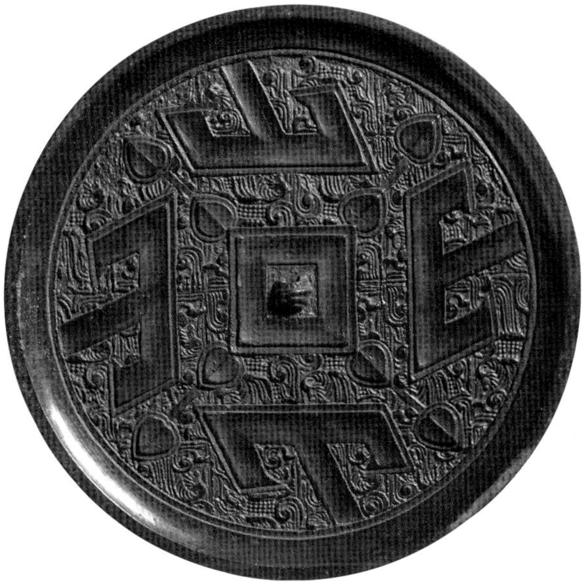

Endowed with a lustrous *heiqigu* (antique black lacquer) patination, this small mirror was exquisitely cast. It has four T-shapes (or *shan*-shape in Chinese) embedded atop a feather-hook ground pattern. The decoration is intensified by the sharp tilt of the bold T-shapes, while their smooth, polished surfaces provide a striking contrast to the gritty background. The horizontal bars of the T's form a square that corresponds to the central square band. At the latter's corner emerge four prongs of twin leaves, which reach to the inner edge of the beveled rim.

This mirror continues the tradition of the dual structure of a ground pattern and superimposed motifs. In fact, one may still be able to observe the seams in the stamped feather-hook design hidden behind the foreground images, just as in the *Mirror with Quatrefoil and Feather Pattern* (no. 1). The twin leaves are but an expansion of the previous mirror, now being made to work in a different, though equally constructive, manner.

Comparable examples can be easily found across a vast territory, but particularly in the Chu region of Hunan and Hubei.[1] From a Warring States burial site in Changsha in the Hunan province comes a mirror that displays identical treatment of the surface motifs.[2] As Sichuan is not far away, the presence of this mirror type should not be a surprise.[3] Although rarely found in archaeological sites, regions such as Xi'an in Shaanxi province, and Luoyang in Henan also provide evidence of its spread.[4] Even the distant Jilin in the northeast, and Guangzhou in the extreme south, felt the influence of these mirrors.[5] We can surmise from this and other data that this mirror type enjoyed immense popularity during the early years, and lingered beyond the Warring States period. Thus a Western Han burial site in Sanmenxia, Henan, yields a specimen with a design essentially unchanged.[6] And the Guangzhou finds also allow us to point to Early Han burials.[7]

Besides the constancy and persistence of the "four T's" tradition, there are also "five T's" and "six T's" traditions. A variety of arrangements are evident, as skillful craftsmen continually discovered new possibilities. Subsequently, the multilevel framework formed by the different arrangements of T-shapes may have foreshadowed the next type of mirror, one with overlapping lozenges (see no. 3).

1. See Deng 1999, 48–60.
2. Hunan 1960, 39, pl. 13.
3. Sichuan/Chongqing 1960, pl. 2.
4. Shaanxi 1959, pl. 15. For Luoyang finds and the discovery of molds for this type of mirror in Yixian, Hebei, see Liu Y 1985, 1010.
5. For the Jilin mirror, see Kong 1992, 40; for those found in Guangzhou, see Guangzhou 1981, vol. I, figs. 90-(1), 90-(2); vol. II, pl. XXXVII-2.
6. See, for instance, Sanmenxia 1996, 6–15, fig. 11.
7. See n. 5 above, relative to Guangzhou finds.

3.

五花折疊式菱紋鏡

Mirror with Five Blossoms and Overlapping Lozenges

Warring States (475–221 BC), 4th century BC

Weight: 105 g

Diameter: 11.4 cm

Rim height: 0.4 cm

Overall height (rim): 0.4 cm

1995.282

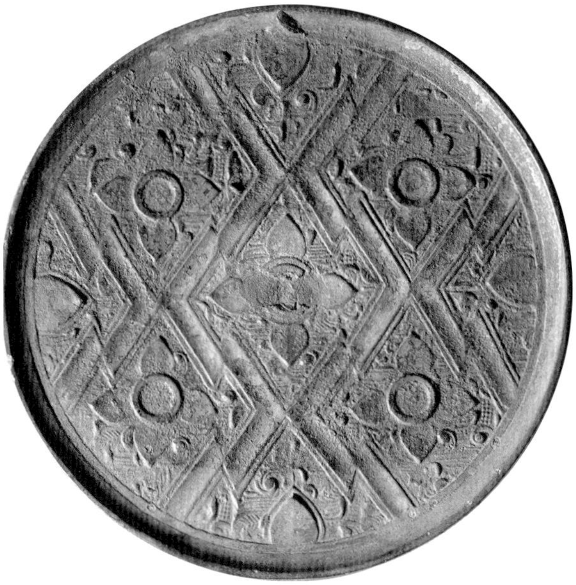

This mirror has a light, cast design of a central blossom surrounded by folded or zigzag bands that resemble, loosely speaking, overlapping lozenges. Within each lozenge, the central blossom is composed of a raised, circular ring from which emerge four petals. Again, a feather-hook pattern similar to that in *Mirror with Quatrefoil and Feather Pattern* (no. 1) is visible underneath the blossoms. Altogether there are five complete lozenges and four incomplete ones. From the outside edge inward, four leaf motifs are placed in those half-lozenges or triangles and aimed toward the knob, which is unfortunately lost.

Though appearing to hark back to mirror no. 1, this mirror also is indebted to the legacy of mirrors with T-shapes (see no. 2). Here the T's have been forsaken, but the use of broad, polished bands with raised edges and light indentations in the middle remains characteristic. They too effectively create bold divisions and images.

Mirrors similar to this blossom-motif one were found in several tomb sites in Hunan province.[1] Another related mirror was found in Sichuan province.[2] The Hubei Provincial Museum has a sumptuous specimen, with evidence of turquoise inset in the center of the blossoms.[3] For a general dating of this type, the inclination has been to place it in the latter half of the Warring States period.[4] Given that this type of mirror came on the heels of the success of the "T" types, the dating appears to be appropriate.

1. Gao 1991, 47–48.
2. Sichuan/Qingchuan 1982, 19.
3. See Deng 1999, 55, fig. 43. Liu Y 1985, 1011, also mentions a similar mirror from Shandong.
4. See Lei 1982, 28 (chart).

4.

透雕紋鏡

Double-tier Mirror

Warring States (475–221 BC), 4th century BC

Weight: 110 g

Diameter: 7.6 cm

Rim height: 0.5 cm

Overall height (knob): 0.8 cm

1999.220

(Plate 1)

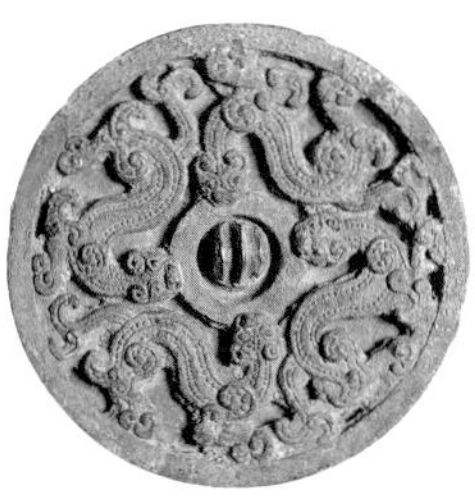

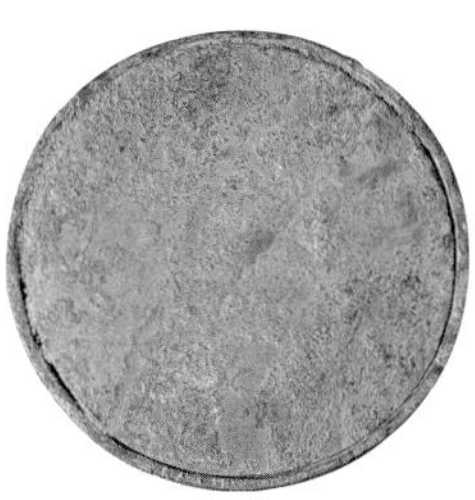

This tiny round mirror is cast in two different layers. The first layer consists of a reflective disc, with a triple-fluted knob cast at its back. Its dimension is slightly smaller than its more elaborate counterpart. In addition to an openwork grille, the second layer features a central ring to allow the knob of the reflector layer to pass through its opening, and its edge folds downward to clasp the full disc in place. The reflecting disc is cast with a high-tin, low-lead alloy, whereas the back of the mirror has an alloy of low tin and high lead—a testimonial to the sophistication in metallurgy in meeting differing requirements.

The small circular ring is the core of the design, to which four serpent-dragons attach their jaws. Their sinuous bodies undulate to enforce a counterclockwise movement. With powerful front and hind legs, each dragon glides on the inner edge of the rim with three-taloned claws visible. Small beads and strong contours outline the body, and tiny horns curl inward on top of the head. The eyes of each creature seem to intensify as its potent snout clenches the central ring.

Excavated examples of such double cast mirrors are few. They can be found in major sites such as Luoyang as well as the Chu region, with Warring States dates.[1] It used to be an article of faith that circular mirrors of this type appeared later than the square ones, but this has not gained universal consensus.[2] More recently, Diane O'Donoghue proposed that this mirror resulted from the combination of a reflector and a circular style of bridle frontlet.[3] Should that be the case, the square mirror essentially has lost its chronological advantage.

1. For Luoyang finds, see Luoyang 1996a, 40; for a Changsha find, in which the mirror is square in shape, see Kong 1992, explanatory note relating to mirror on p. 144.

2. See Lei 1982, 28 (chart), where he places the square double-tier mirrors starting from late Chunqiu era to the end of the Warring States, and the circular ones in the second half of the latter period.

3. See O'Donoghue 1990, chap. 4, 63ff.

5.

蟠螭菱紋鏡

Mirror with Serpentine Interlaces and Angular Meanders

Warring States (475–221 BC), late 4th–early 3rd century BC

Weight: 258 g

Diameter: 16.9 cm

Rim height: 0.5 cm

Overall height (knob): 1 cm

1995.270

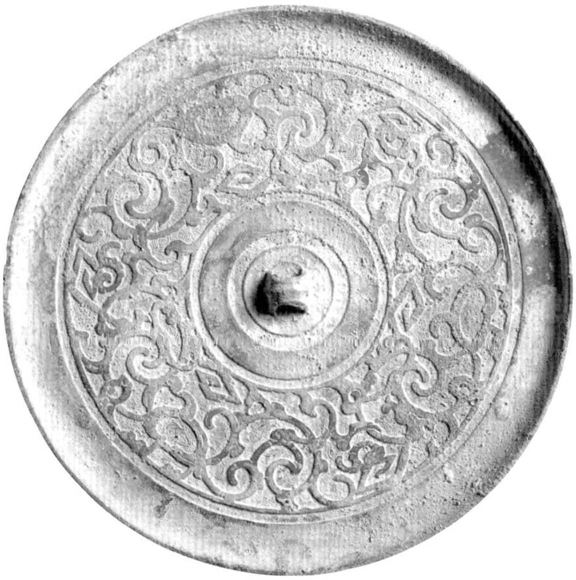

This mirror is essentially organized around a triple-fluted knob within a series of concentric circles. A low relief of serpentine interlaces encircles the middle band. Three heads are visible, with their bodies coiling around each other in an endless curling and uncurling movement. Each quadruped is formed with thin stripes and occasional lineal indentations to differentiate body parts. A set of three modest and angular meanders appears in the inner edge of the band; another set of similar shapes is more widely spaced in the outer register. This is a harbinger of the much more pronounced treatment in mirror no. 6. A ground decoration of "curls and striations," lacking in clarity and distinction for the most part, continues underneath the interlaced serpents.

Mirrors such as this enjoyed wide distribution. Regions from the Henan to the east bear witness to its popularity.[1] An example is that found in the Henan region, with four zoomorphs encircling the central band, connected by continuous arabesque scrolls.[2] The Chu region too has yielded a considerable number.[3] A mirror found in Banbishan, Yangxin, in Hubei province also resembles the Carter example, dating to the middle phase of the Warring States period.[4] Such early mirrors lack even the modest beginning of the lozenge shape. By and large, this feature received increasing emphasis as time went on; it can be found in early Western Han tombs in Shandong, Hunan, and Hubei. At least some of these may still be of Warring States pedigree.[5]

1. See Liu Y 1985, 1009–12.

2. Zhu 1988, 472.

3. See Kong 1992, 103–4, 106; all three mirrors hailed from Changsha, Hunan.

4. Xianning/Yangxin 1994, 529.

5. See Kong 1992, 122, 125, 126, respectively.

6.

蟠螭菱紋鏡
Mirror with Serpentine Interlaces and Angular Meanders

Warring States (475–221 BC),
3rd century BC
Weight: 371 g
Diameter: 17 cm
Rim height: 0.7 cm
Overall height (knob): 0.9 cm
1995.277

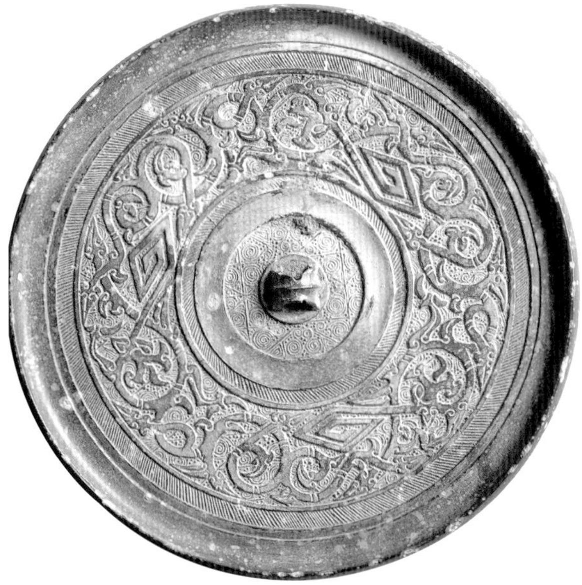

Three intertwined creatures swirl around the central circular band of this crisply cast mirror. Intense eyes, massive snouts, and protruding tongues display their power. Bodies curve in upon themselves, attenuated forelegs pointing with sharp talons. Double stripes give a sense of roundness to the slender torsos, and other curved lines mark, and differentiate, the twist and turn of their bodily parts. The ground pattern of whorls and granulation persists underneath them as well as around the triple-fluted central knob. Amid the concentric rings that form the basic principle of design, two hachured bands frame the interlacing arabesques.

Noteworthy is the decisive role played by the angular meander motif, a step beyond that in mirror no. 5. Upon close observation, these angular forms appear to be a whimsical extension of the creatures' antlers. The prominence of this motif provides a clue to the possibility that the mirror may be dated to the last phase of the Warring States period or perhaps to as late as the early Western Han dynasty. A mirror from an early Western Han burial in Changsha shares the same structural kinship, and helps to buttress this dating.[1]

In the present state, the mirror's surface has a satiny patina with light buff encrustation.

1. Kong 1992, 125.

7.

三龍鏡
Mirror with Three Dragons

Warring States (475–221 BC),
3rd century BC
Weight: 249 g
Diameter: 16.9 cm
Rim height: 0.45 cm
Overall height (knob): 0.6 cm
1995.280

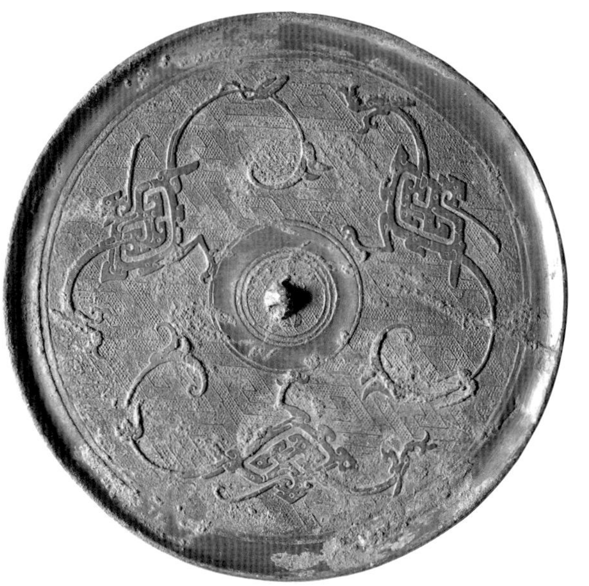

This mirror has a thin strip or beveled border, with restricted, concentric rings in the center. This arrangement, which leaves a broad space to accommodate a granulated and textured ground, in turn offsets the form of the three dragons. Cast in low relief, the dragons revolve around the knob in a counterclockwise motion. Each has a modified-lozenge body and stands on one leg with sharp talons attached to the inner register surrounding the small knob. Emerging from the lozenge body is a long, slender neck on one side, and an even longer tail on the other. Attached to the neck is a small head with open jaws, as if trying to catch the next dragon's tail—a tail executed in a sweeping and sinuous curve with extra appendages and curled hooks.

One notable feature of this mirror is that the dragons are seen in a centripetal mode, in contrast to the characteristic centrifugal mode in previous examples. This perceptual inversion brings about a different way of organizing the decoration on the mirror's back and is effective in broadening the range of artistic exploration.

Long known and well studied, this type of mirror had its origin in the Warring States period and extended into the Western Han period. One may refer to a specimen found in the royal tomb of the second King of Nanyue, ca. 122 BC in Guangzhou.[1] This Nanyue example was likely imported from without, a distance factor that further delayed its final interment. A date of late Warring States can be assured for this Carter mirror, if not actually earlier.

1. Quan 1998, 314, fig. 5-(2).

8.

雙花龍鳳鏡

Mirror with Double Quatrefoils, Dragons, and Phoenixes

Warring States (475–221 BC), 3rd century BC
Weight: 805 g
Diameter: 23.2 cm
Rim height: 0.5 cm
Overall height (knob): 1.2 cm
1995.279

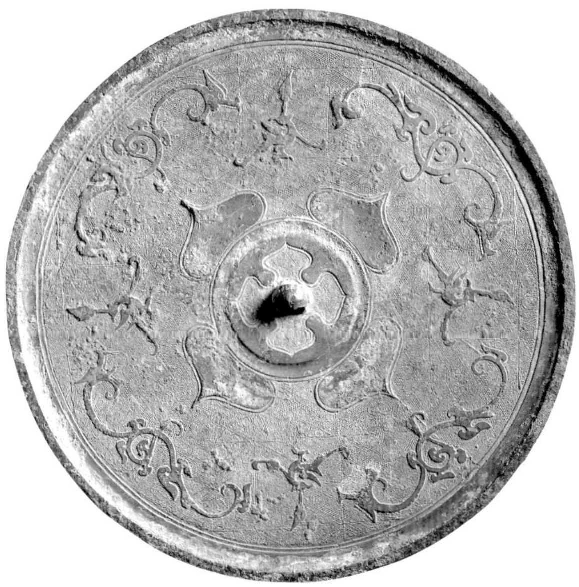

Fine repair work has restored the integrity of this large mirror, where once it was broken into nine fragments. It continues and expands the thematic evolution begun in *Mirror with Three Dragons* (no. 7). Now there are four dragons and four phoenixes revolving around a small, domed knob surrounded by double quatrefoils. Like the previous mirror, the granulated diaper pattern of the ground is given an expanded share of the mirror's back. The dragons are proportionally smaller in relation to the mirror's enlarged dimension, their bodies further attenuated and made slender as they lose the lozenge-shaped torso and the clarity of anatomical details. Precisely because of this, their simpler constitution contributes to a curvilinear fluency in style. In between the dragons, the phoenixes claim a far more modest space, with spreading wings and a tall crest. The arrangement is by no means static as the double quatrefoils expand and contract, and the zoomorphs curl down and ascend. Again, as in mirror no. 7, the viewing mode is centripetally oriented.

9.

雲雷紋地連弧紋鏡

Mirror with Continuous Arcs against Whorl Pattern

Warring States (475–221 BC), 3rd century BC
Weight: 557 g
Diameter: 23.7 cm
Rim height: 0.75 cm
Overall height (knob): 1.2 cm
1995.278

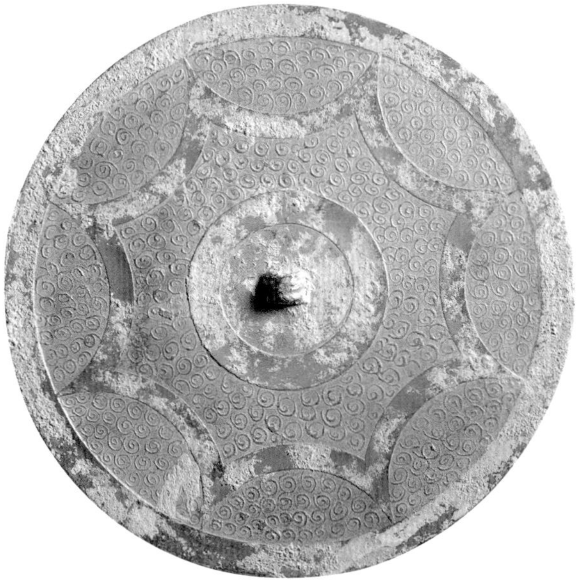

This flat but large mirror has a plain rim enclosing eight linked arcs, which form an octafoil or sunburst. A whorl pattern covers the ground, and double rings surround the triple-fluted knob. The design is simple, but bold and effective. The continuous-arc band is an important motif that will be repeated later in a number of mirror types.

Comparable mirrors have been found in the Chu region, dating to the Warring States period.[1] Without doubt, this type of mirror prevailed during the late Warring States and lasted into the Western Han period. In Chengdu, Sichuan province, an early Han burial site yielded a mirror that was analogous in design, except for its smaller size and a reduction of one of the linked arcs.[2] In remote Nanyue, three versions of this design were the cherished objects in the royal burial (122 BC).[3]

1. See Li ZG 1957, pls. 20-(5), 20-(6). See also Lei 1982, 28; Lei places the emergence of this typology in late Warring States, continuing into the Qin and early Western Han periods. For potential variations of this type, see Kong 1992, 132–40.

2. Sichuan/Chongqing 1960, pl. 11.

3. See Quan 1998, 308, also pls. 3-(9), 4-(6), 5-(1).

Phase II

Han and Post-Han Dynasties

206 BC–AD 589

10.

方格四乳鏡

Mirror with a Square Band and Four Nipples

Western Han (206 BC–AD 8), late 3rd–early 2nd century BC
Weight: 62 g
Diameter: 8.4 cm
Rim height: 0.35 cm
Overall height (knob): 0.7 cm
1995.275

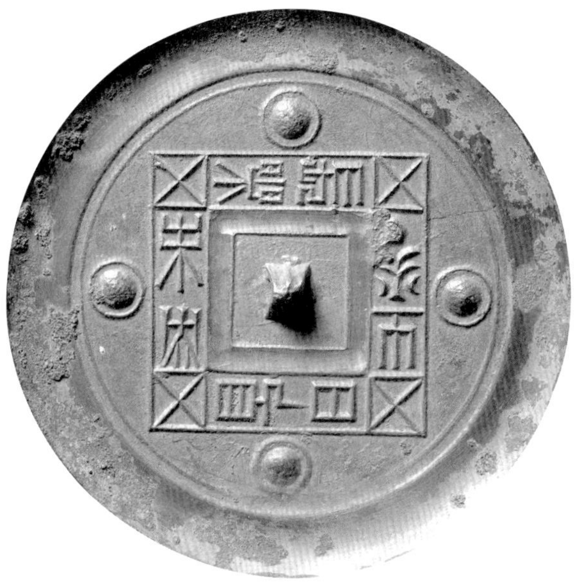

This tiny mirror has a triple-fluted knob, which looks like a folded bridge. It rises from a square base and assumes a triangular profile. Immediately adjacent to the base is another square band with an inscription surrounding the central complex. Beyond the square band, four nipples take their rightful place at axial points. A beveled rim displays its upturned edge and completes the design.

The square band with eight characters cast in thread-relief dominates the mirror. Each side of the square contains two characters, and a simple cross marks the corner junctions. The seal script of the inscription is noted for its straight lines and rigid formation—a fitting match for the simple and stark design. The inscription reads: 見日之光，長樂未央 (May you see the light of the sun, and live in endless joy!).

So far, archaeological excavations have unearthed no mirrors that exactly match this Carter version, although a number of mirrors from Xi'an in Shaanxi province share the basic platform.[1] Like the Carter version, they are small thin discs with beveled edges, a square in the center, and four nipples at axial points. However, some minor changes or shifts always are present to undermine the resemblance. The inscription may be identical, but it occupies a different place. The nipples may overlap the square to allow for extra touches. It is possible that the stark design may not have appealed to clients, and the mirrorsmith was compelled to opt for more elaborate treatment. As a result, the primal object, such as the Carter mirror, vanished under the accumulated series of popular variations. Eventually there emerged a larger and far heavier piece, which had the same features of a square band with inscription and four axial nipples. But this newer version replaced the odd knob with a winsome, hemispherical dome, and the beveled rim became a considerably thickened edge outlined by sixteen linked arcs. Also added were pairs of grass and other plant motifs alongside the nipples (see no. 11).

1. Shaanxi 1959, pls. 6–8.

11.

方格四乳草葉鏡

Mirror with a Square Band, Four Nipples, and Grass Leaf Motifs

Western Han (206 BC–AD 8), 2nd century BC
Weight: 809 g
Diameter: 18.3 cm
Rim height: 0.55 cm
Overall height (knob): 1.2 cm
1995.285

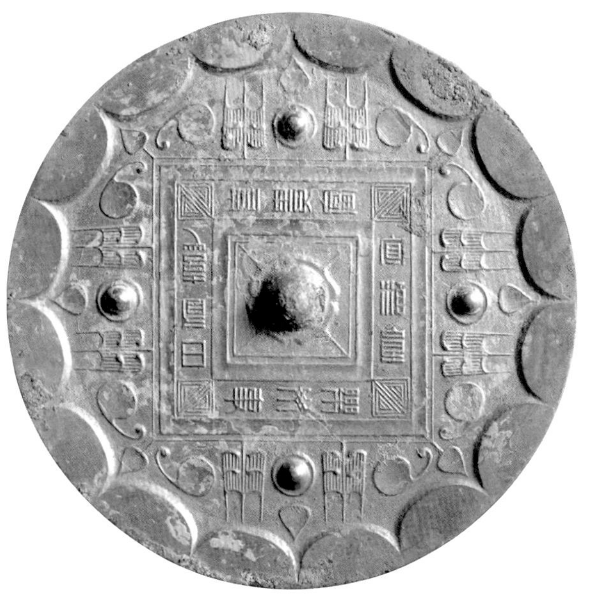

Having already described the essential characteristics of this type of mirror (see no. 10), it may be equally important to underscore its popularity. After all, the square-in-the-circle modality and the linked arcs that evoke the sunburst are two of the most enduring motifs in the history of the Chinese mirror. In the former aspect, the Carter mirror and those like it can be seen as a precursor to the TLV mirror and, at a given point, also merged with it.[1] In the latter aspect, the mirror explores the sunburst image with the same zeal as other popular categories of mirrors known as *riguang jing* ("sunlight" mirror; see no. 14) and *zhaoming jing* ("illumination" mirror; see no. 15).

A full tabulation of the evolution of the mirror typology has been only partially completed. In the latest investigation, Bai Yunxiang counted more than forty-six pieces of this type of mirror with the "older" inscription of 見日之光，長樂未央 (May you see the light of the sun, and live in endless joy!).[2] Those with the "newer" version of the inscription, such as in this mirror, will have to await further scrutiny. Contrary to the emphasis on sunlight, the Carter mirror offers an invocation of general blessings, with good fortune and joyful plenitude for humanity:

長貴富，
宜酒食，
樂毋事，
日有熹。

May you enjoy lasting fortune.
May you enjoy wine and food.
May you enjoy freedom from worry.
May your days be replete with happiness.

Another way to gauge the popularity of a particular type of mirror is by the extent of distribution. Thus we can find representative examples in the centers of the Han empire, such as Chang'an and Luoyang.[3] We can also spot them in Shanghai on the coast, or south of Shanghai in Anji, Zhejiang.[4] Most important is that this type of mirror appears as a part of funerary furnishing in the tomb of Liu Sheng 劉勝, the Prince Jing 靖 of Zhongshan 中山, who died in 113 BC and was buried in Mancheng, Hebei.[5] It is perhaps the best dated example for the type, even though this particular mirror must have been made a couple of decades before its final burial. Having an illustrious patron such as Prince Jing contributed immensely to the rising prestige of the mirror itself. Elsewhere, Hunan yielded three examples and Sichuan alone unearthed a total of ten, the majority from one site in Chengdu.[6]

1. See for instance the mirrors reproduced in Kong 1992, 203, 205.
2. See Bai 1999, 65.
3. Shaanxi 1959, pls. 9, 10; Luoyang 1988, pls. 5, 6.
4. See Wang Z 1988, 694–717, fig. 11; Anji 1996, 56, fig. 12 (upper right). This mirror is estimated to date from a period after 118 to about 88 BC.
5. Hebei 1996, pl. 20.
6. See Hunan 1960, pls. 49–51; Sichuan/Chongqing 1960, pls. 13–22.

12.

四花連弧鏡
Mirror with Quatrefoils and Linked Arcs
Western Han (206 BC–AD 8), 2nd century BC
Weight: 234.5 g
Diameter: 14 cm
Rim height: 0.6 cm
Overall height (knob): 0.6 cm
1995.276

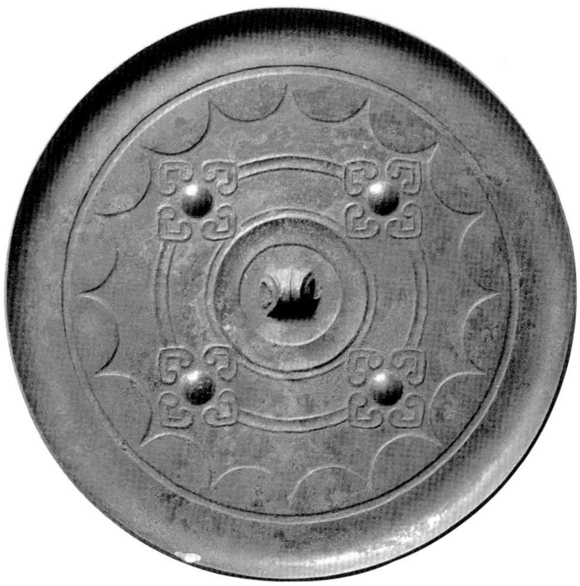

This mirror exhibits the following earlier traits: it is thin, light, and has a beveled rim. The triple-fluted knob is also an archaic feature. Otherwise, the design is a series of concentric circles, with a row of linked arcs at the outer register before the rim begins—a motif that would increase in popularity. Four nipples, each on a quatrefoil base, establish the semblance of a square. They are the principal accents, relieving the design from a simple, even subtly executed, layout.

The importance of this mirror is twofold. First, it stands as the beginning of a new type. A mirror that was found in a Guangzhou site of the early Western Han period is perhaps the closest of all in appearance to this Carter version.[1] Second, it is rare, and its rarity is a combined result of its own primal state and the subsequent production of more ornate versions, which overwhelmed it in quantity as well as in survival rate. An example is the far larger, and by all means the most richly decorated, specimen known to us.[2] It comes from the 104 BC burial of Dou Wan 竇綰, who was the wife of Liu Sheng, the previously mentioned Prince Jing of Zhongshan.

1. Mo 1958, 39ff, pl. 14-(6).

2. See Hebei 1996, color pl. 4, pl. 17.

13.

星雲鏡

Mirror with Clouds and Nebulae

Western Han (206 BC–AD 8), 2nd century BC

Weight: 185 g

Diameter: 10 cm

Rim height: 0.4 cm

Overall height (knob): 1.2 cm

1995.288

(Plate 2)

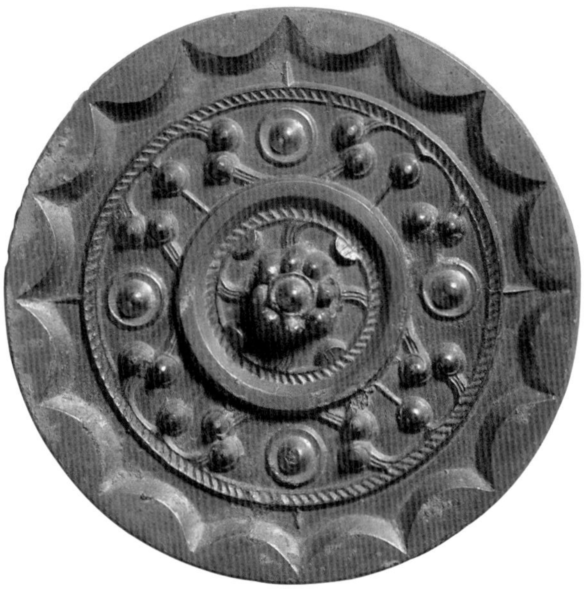

Despite its title, it is unlikely that "cloud and nebula" was the intended theme for a mirror of this type. This mirror's back offers a concentric series of three distinct zones. The first zone contains the knob plus its immediate environment, outlined by a raised ring. The knob itself is raised from its base to an eightfold fluted column, culminating at the top with a central apex surrounded by eight lesser ones. The second zone is the inner band, where four larger nipples on circular bases are the principal accents. Over each larger nipple is a curved arm with two smaller nipples at its ends. In between the larger nipples are also three smaller nipples linked by a U-shaped scaffold. Thus is created an alternating sequence of curves and reversals, while the sets of lesser nipples also set off a studied rhythm of linkage and separation. The last zone displays sixteen linked arcs at the rim, thickened to lend a further sense of substance and weight. Indeed, the whole mirror projects a sculptural impression and appears crisp and tectonic. In that sense, it is easily recognizable as a separate category amid Han mirrors.

This type of mirror is found in many areas in China. Right at the center of the empire, in the city of Xi'an, archaeological excavations have unearthed a number of examples, with one that is fairly close to the Carter version in design and execution.[1] Luoyang also supplies at least two fine examples from Western Han tombs.[2] While Hengyang in Hunan has unearthed a variant of the "Clouds and Nebulae" mirror, finds in Chengdu in the province of Sichuan comprise more varieties.[3] We can also broaden our coverage to Shanxi, Hebei, Shandong, and then southward to Shanghai.[4] Even the remote southern provinces of Yunnan and Guizhou yield comparable pieces.[5] Most of the burial sites are dated to the mid and late Western Han period.

1. See Shaanxi 1959, pls. 15–17, especially pl. 16.

2. See Luoyang 1988, pls. 8, 9; pl. 8 is identical to the Carter version.

3. See Hengyang 1993, 243; Sichuan/Chongqing 1960, pls. 28, 30.

4. See Pingshuo 1988, 446, fig. 9; Hebei 1990, 324, fig. 3-(7); Zhucheng 1987, 782, fig. 6; Wang Z 1988, 704, figs. 1, 4.

5. See Guizhou 1986, 199ff, fig. 21-(7). For Yunnan examples, see Chen B 1982, 332.

14.

重圈連弧銘紋鏡

Mirror with Concentric Circles and Linked Arcs
Western Han
(206 BC–AD 8),
late 2nd–1st century BC
Weight: 132 g
Diameter: 8.1 cm
Rim height: 0.5 cm
Overall height (knob): 0.9 cm
1995.290

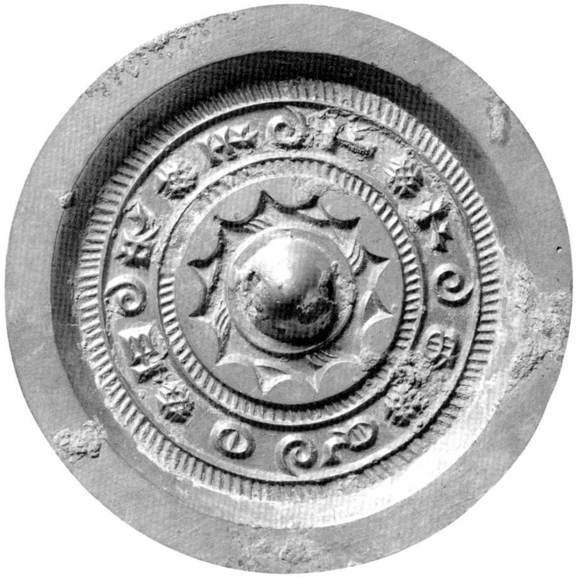

In this modest mirror, the linked arcs around the hemispherical central knob create a sunburst effect. Outside that central core, two hachured rings enclose a middle band reserved for an inscription. A plain, but thickened, rim completes the mirror design.

The inscription contains only eight characters. Two alternate types of spacers are inserted between each word. The written content enforces the sunburst motif and intensifies the impact:

見日之光，天下大明。

By the light of the sun,
The world is made bright.

This is the so-called *riguang* 日光 (sunlight) mirror, as it is often designated in China by virtue of the inscription's first line. This designation is intended to distinguish it from such others as the *zhaoming* 昭明 (illumination) mirror, the *tonghua* 銅華 (copper essence) mirror, and the *qingming* 清明 (limpid purity) mirror. All these types of mirrors have a basic kinship in design, except for the fact that the first line of the inscription contains differing characters and aims toward differing contents. Since most appellations for mirror types are design-based, introducing inscriptions as criteria for classification only creates needless confusion. Notable examples are when a mirrorsmith decided to combine inscriptions from both the sunlight and illumination mirrors, or from both the sunlight and limpid purity mirrors.[1] A better strategy would be to consider these as subcategories of one and the same generic type, with the sunlight mirror as perhaps the simplest and therefore also the earliest to appear.[2]

That being the case, it is difficult not to note the modest dimension and style for this Carter mirror. Even the calligraphy, cast in relief, is not well regulated—with some characters in strokes resembling the cuneiform, while others are more in line with the seal script's characteristic curvilinear lines. Regardless, as a mirror it cannot be ignored. Its modest style and, presumably, price enabled it to reach a broad clientele. Archaeologically, this subtype enjoyed a fairly extensive distribution, roaming from the center of the empire all the way to peripheral regions such as the west, northeast, and extreme south. Most sunlight mirrors are quite similar in appearance, regardless of whether they hailed from Xi'an,[3] Luoyang,[4] or Liaoning.[5] The Carter mirror is a typical example.

1. For combining the inscriptions from the "sunlight" and "illumination" mirrors, see Anji 1996, 56, fig. 12, lower left; for combining the inscriptions from "illumination" and "limpid purity" mirrors, see Chongqing 1986, 238, fig. 9.

2. Note the confusion so created in an archaeological report, where several different types of mirrors are being designated as sunlight mirrors because of their identical inscriptions. See Shunhua 1983, 851, especially figs. 1-(3), 2-(3). Both feature a square-in-circle pattern, and the latter is identical to Carter mirror no. 11.

3. See Shaanxi 1959, pls. 34, 35. The mirror in pl. 34 has a beveled rim, whereas that in pl. 35 shares with the Carter mirror the same thickened, plain rim.

4. See Luoyang 1988, pls. 18, 20.

5. For Liaoning finds, see Yu 1958, 71–90, fig. 17-(4).

15.

重圈連弧鏡

Mirror with Concentric Circles and Linked Arcs

Western Han (206 BC–AD 8), 1st century BC
Weight: 421 g
Diameter: 11.9 cm
Rim height: 0.75 cm
Overall height (knob): 1.1 cm
1995.291

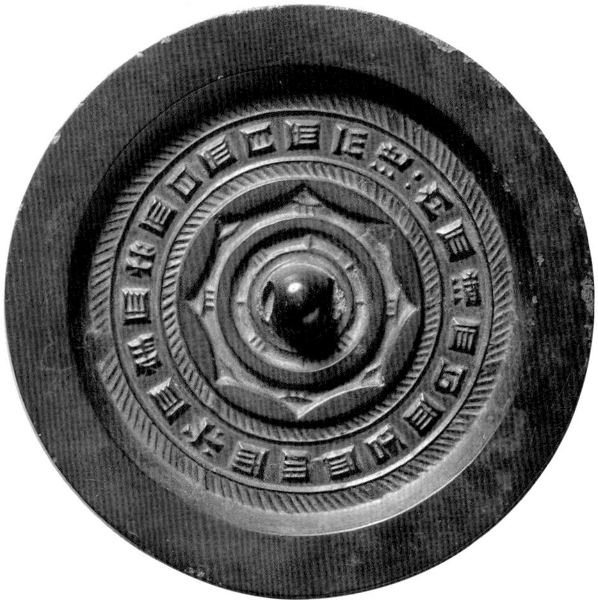

This mirror is the "illumination" mirror—a fairly typical one at that. It has inherited much from the "sunlight" mirror, with a hemispherical knob at the center and a sunburst around it. A thickened, plain rim leads to a deeper inner well. A band of inscription is at the middle register. The inscription is deceptively long, as the spacer 而 appears repeatedly between words, which doubles the length. The mirrorsmith miscalculated, and at the end had to shorten the text to the point where the last line no longer is meaningful. Below, we shall translate it by incorporating the full text:

內清以昭明，
光象夫日月，
[心忽楊而愿忠，然雍塞而]
不泄。

Its inner purity is shown in perfect illumination.
Its light is the image of the sun and moon.
[My heart experiences uplift and is keen to be loyal,
Yet it is obstructed and] unable to express it.[1]

Tolerance for such miscalculation was high. As with the spacer, the characters are now lined neatly in order, each word a framed rectangle. The immediate impression is one of substance and stability. It can be said that the interior remains fluid, with circular bands revolving in either clockwise or counterclockwise manner.

The "illumination" mirror is found across China, often appearing side by side with the sunlight mirror in given sites. It is possible to speculate that, after its introduction, the illumination mirror began to overshadow the older type and moved on to dominate the market. Instead of surveying the entire distribution network, we will simply identify a few sites in which those similar or identical to the Carter version are found. Along this line, we can identify a Luoyang specimen which, even though smaller in diameter, is as close to the Carter version as any.[2] Even the calligraphy and the spacer are identical. So is the wide border, which lends a sense of weight. In Hebei province, a mirror resembling the Carter version can be dated to the late Western Han period.[3] It, too, is smaller, but in inscription is identical, even in missed words. In Hunan, eight such mirrors have been unearthed. Of varying sizes and reflecting a subtle degree of change in border width, these are nevertheless consistent in the mode of presentation.[4] All are dated to late Western Han by virtue of their burial sites. This only suggests that illumination mirrors may well have had an earlier beginning. In terms of this lower limit, some of the illumination mirrors were found in Xin era burials; others crossed over to the Eastern Han.[5]

1.
The translator notes that the Han mirrorsmith often simplifies, abbreviates, and reduces the inscription text, sometimes to a degree that is no longer comprehensible. For full passage of this inscription, see Kong/Liu 1994, 55.

2. Luoyang 1988, pl. 21.

3. Hebei/Zhangjiakou 1990, 12, fig. 21-(1) (upper left).

4. See Zhou SR 1986a, 116–17, figs. 34–41.

5. Ibid., figs. 48, 49.

16.

重圈連弧鏡

Mirror with Concentric Circles and Linked Arcs

Western Han
(206 BC–AD 8),
1st century BC
Weight: 566 g
Diameter: 14.9 cm
Rim height:
0.7 cm
Overall height
(knob): 1.1 cm
1995.306

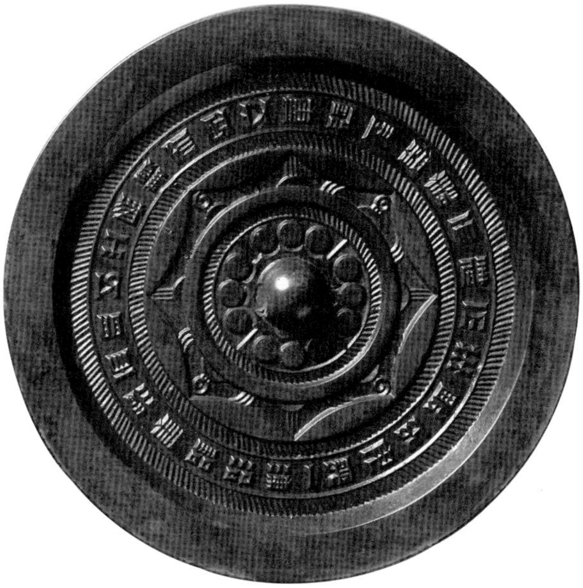

If we were to define mirror types by the first line of inscription, this would have to be called a *tonghua* (copper essence) mirror. Such is the folly of attempting a classification system based on this approach. Thus the mirror declares:

湅治銅華清而明。
以之為鏡宜文章。
延年益壽而辟不羊(祥)。
興天毋極。

Refine the copper to be pure and bright.
Make it into a mirror to receive rightful decor.
May it extend your life and remove misfortunes!
[You are] coeval with the heaven above.
[Like the sunlight, shining forever].[1]

In terms of design, this mirror is just one more step beyond the previous illumination mirror (no. 15). Its casting is crisp and sharp. It is also larger in dimension, appearing tectonic and substantial. The *heiqigu or* antique black lacquer patination also enhances this impression. Within the innermost circle, a set of twelve small round discs surrounds the hemispherical knob, a feature already known in some illumination mirrors.[2] Whatever changes there are consist of minor tinkering, such as an increase in the number of concentric circles and hachured ribbons. The sunburst remains a key feature to link several subtypes, as does the strong plain border.

Because the copper essence mirror is a minority type within the larger genus, it is difficult to investigate these mirrors in an isolated manner. It is best to fuse them with the limpid purity mirrors and to regard them as the last flowering of the broader category of sunburst mirrors, dated toward the last phase of the Western Han period. One of the finest of the limpid purity mirrors is that found in Dingxian, Hebei province, as a part of the tomb furnishings for Liu Xiu 劉修, Prince of Huai, who died in 55 BC.[3] This also confirms its time frame.

1. Spacers are not used here. One way to complete the full text is, after the above, to add the following: 如日之光，長未央。See Kong/Liu 1994, 55.

2. See Luoyang 1988, pl. 22.

3. Hebei 1996, pl. 36.

17.

四葉重圈連弧紋鏡

Mirror with Quatrefoil, Concentric Circles, and Linked Arcs

Eastern Han
(25–220),
late 1st century AD
Weight: 406 g
Diameter: 15 cm
Rim height: 0.5 cm
Overall height
(knob): 1.2 cm
1995.318

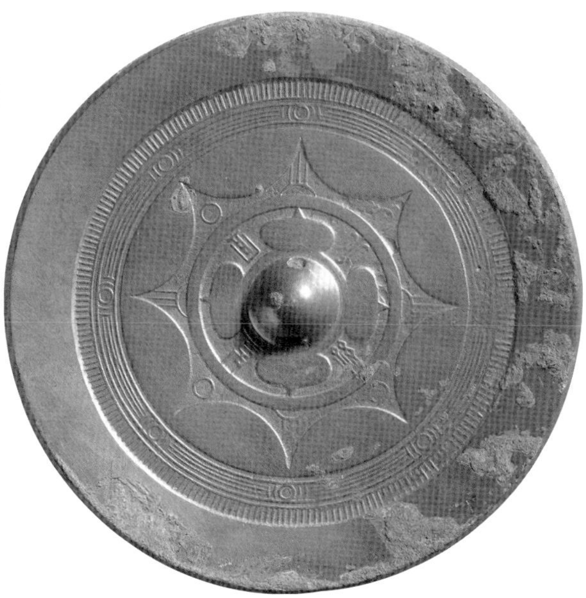

A simple and effective design, this mirror has a hemispherical knob surrounded by a series of concentric circles. From the knob radiate four leaves, with inscribed words interspersed in between: 長宜子孫 (May [the mirror be] cherished by your sons and grandsons for a long time). A ring then encloses this. The second circular band contains a series of linked arcs to form a sunburst pattern; between each set of two arcs alternate two minor motifs—a small circle and a simple arching mound with three uprights. The band that encloses the sunburst utilizes a series of six-string courses, divided by eight small circles and double parentheses, vaguely reminiscent of the egg-and-dart motif. Beyond that, a simple ring of hachured lines contrasts with the plain, wide border.

The design is distinguished not only by its linear elements, but also by the subtle interplay of layers and strata. Classical in appearance, this type of mirror is marked by its sunburst core. It descends from a long and involved heritage, and also gives rise to a major branch in mirror evolution (see nos. 34–37).

Dating of this mirror requires an adjustment between the specific typology versus the individual piece. For the type, a Western Han period date is assured in the face of a Guangxi find of third century BC.[1] Other finds include a specimen from a Xin period burial (AD 9–24) in Hanjiang Xian near Yangzhou, Jiangsu province,[2] and another dated by inscription to AD 64.[3] Also, archaeological excavations have unearthed examples from mid or late Eastern Han periods throughout China in provinces such as Shaanxi, Henan, Shandong, Sichuan, and Hunan. The first three mirrors mentioned above set a broad time frame, and the later discoveries obviously extend its wide reach and popularity.

This mirror differs from the earlier ones, including the mirror dated to AD 64, in only one detail: in the area containing the six-string courses, it favors a perfect circle rather than the volute. Also, its inner sets of four strings are straight and parallel instead of displaying, from both ends, a gradual slide and a degree of overlap. The changes here may have deprived the Carter mirror of its prototypes' intricate rhythm, but it has also gained a simple and effective clarity. Quite possibly, this is a feature that will place it in the early years of Eastern Han.

1. Hepu 1995, 226–83, especially p. 229, fig. 4-(1).
2. Yangzhou/Hanjiang 1991, 39ff, fig. 16.
3. Kong 1992, 366.

18.

四乳四烏鏡
Mirror with Four Nipples and Four Birds

Western Han (206 BC–AD 8), 1st century BC
Weight: 93 g
Diameter: 6.2 cm
Rim height: 0.6 cm
Overall height (knob): 0.95 cm
1995.297

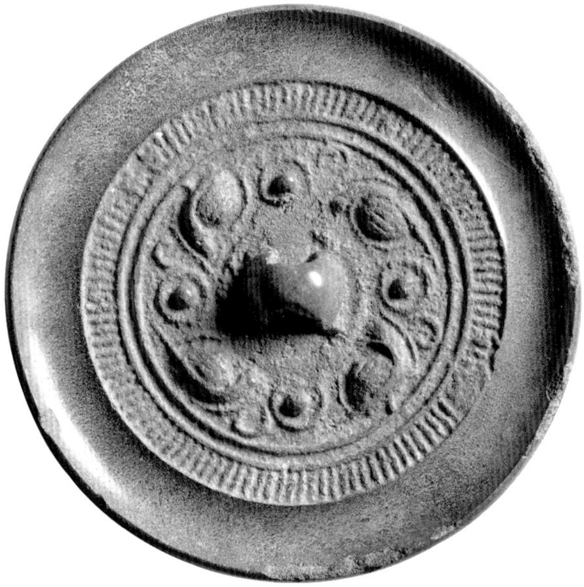

A simple and modest piece, this mirror's charm lies precisely in its simplicity and modesty—in scale and concept. Nothing elaborate is here to confuse the viewer. Even its surface is fretted, and the only smooth and lustrous portion lies in the knob and the border, with its subtle greenish patination. The four nipples are enclosed within circles, and the birds, resembling quail, display soft contours and ripe bodies. The concentric design is modified by thin relief lines and sunken surfaces, and also by the contrasting use of decorative motifs, including a ribbon with hachured lines. The rim curves upward in a smooth flow much like the early mirrors of Western Han.

Lending additional support to the dating is the Echeng find of the Xin era, which has not only the ring of hachured lines, but also an additional sawtooth band.[1] Based on the Carter mirror's utter simplicity, one cannot but be convinced that it preceded the Echeng find by decades.

1. Hubei/Echeng 1986, pl. 9.

19.

四乳八禽鏡

Mirror with Four Nipples and Eight Birds

Western Han (206 BC–AD 8), 1st century BC
Weight: 236 g
Diameter: 9.6 cm
Rim height: 0.65 cm
Overall height (knob): 1 cm
1995.305

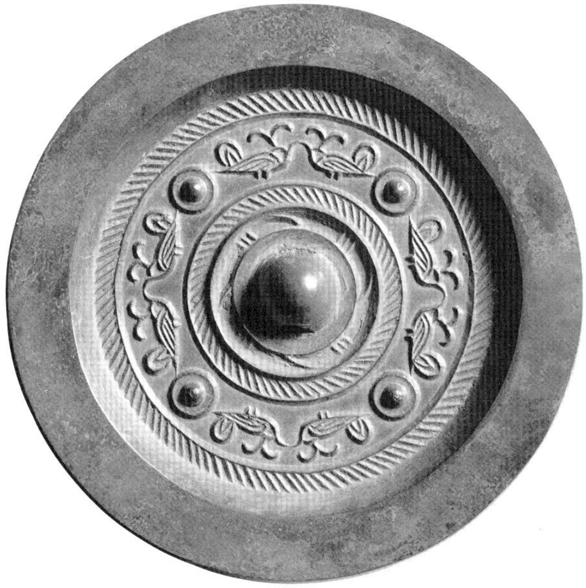

This is a small mirror, but finely and crisply cast. Inside the thickened border, the inner well sinks deep. A hemispherical knob is at the center, surrounded by concentric rings. In the main decorative band, the four nipples form the semblance of a square. Between them are facing pairs of stationary birds, with beaks linked, antenna-like crowns, and uplifted, rounded tails. Accentuating the design are two ribbons of hachured lines in oblique alignment, enforcing a sense of endless revolution.

While not exactly a popular mirror type, its distribution is sufficiently widespread. Luoyang and Xi'an in the north, and Echeng, Yancheng, and Guangzhou in the south have yielded examples.[1] The farthest ranging, perhaps, are the Jilin finds in the northeast.[2] There is little doubt that this type can be dated to the Western Han period.

1. For Luoyang and Xi'an finds, see respectively, Luoyang 1988, pl. 15; and Shaanxi 1959, pl. 54. For Echeng, Yancheng, and Guangzhou finds, see respectively, Hubei/Echeng 1986, pl. 8; Kiangsu/Nanking 1964, 396, fig. 8-(2); and Kong 1992, 263.

2. For Jilin finds, see Kong 1992, 225, 263.

20.

四乳四虺鳥紋鏡

Mirror with Four Nipples, Quasi-Dragons, and Birds

Western Han (206 BC–AD 8), 1st century BC
Weight: 173 g
Diameter: 9.9 cm
Rim height: 0.4 cm
Overall height (knob): 0.9 cm
1995.314

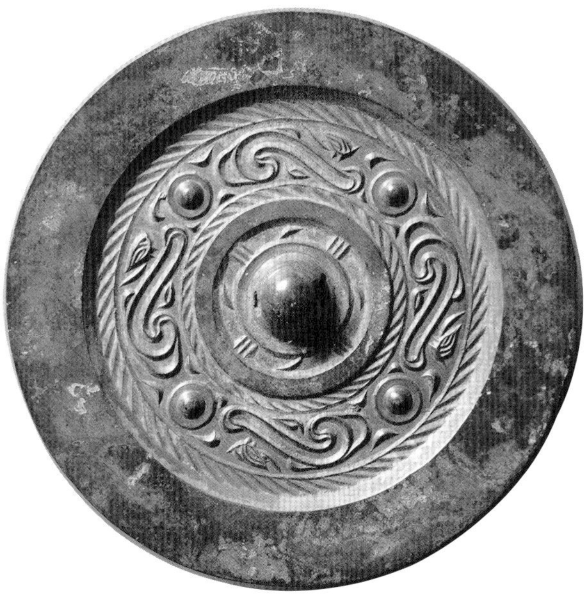

As a mirror type, this one is often known by the title of 四乳四虺 (Four Nipples and Four *Qiu*) or 四乳四螭 (Four Nipples and Four *Chi*). Whether *qiu* or *chi* is immaterial, as both refer to dragon-like creatures. The fact is, the undulating organisms between the nipples cannot be identified as anything but scroll forms, although they do seem to have a will of their own. In response to the suggestion that these organisms may have been debased or mutated into their present forms, we coin the term "quasi-dragon." Small birds show up beneath them, and leaf sprays are above. The knob has a seating, but it is also surrounded by a circular ring, with a cross of three lines to mark the four axes, and simple raised curves in the intervening spaces. A broad, plain rim encloses the whole mirror.

In Donghai Xian, Jiangsu, a burial datable to 10 BC yielded a specimen that is quite close in design.[1] One may naturally assume that the production of that mirror is at least one or two decades earlier, and the evolution of this type of mirror may be earlier still. The Carter mirror should share the same chronological framework.

On the other hand, the mere mention of Jiangsu should not limit the type to the eastern province. In fact, distribution of this type is widespread. Xi'an in Shaanxi, Luoyang in Henan, Shaoxing in Zhejiang, Echeng in Hubei, Chengdu in Sichuan: these are but a few of the major sites that have produced examples from Western Han burials.

1. Lianyungang 1996, 4–31, especially p. 7, fig. 11.

21.

四乳龍虎鏡
Mirror with Four Nipples, Dragons, and Tigers
Western Han
(206 BC–AD 8),
1st century BC
Weight: 572.5 g
Diameter: 4.2 cm
Rim height:
0.7 cm
Overall height
(knob): 1.1 cm
1995.300

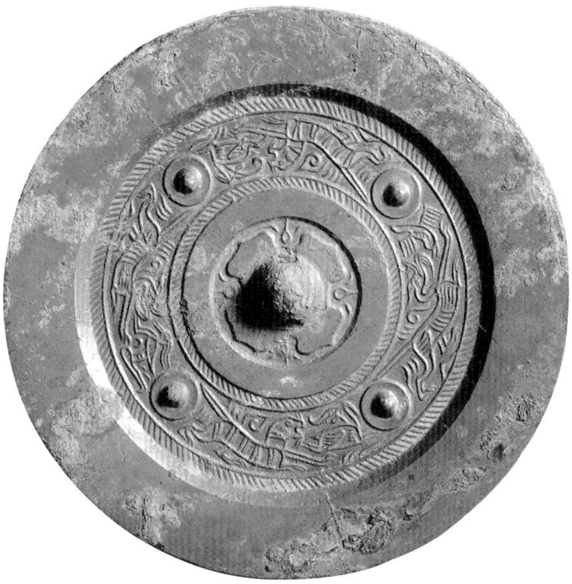

The knob is surrounded by a quatrefoil; a raised, plain circular band and a hachured ring extend the domain. In the inner well, the decorated zone is made narrow and the space defined by nipples appears to be stretched and lengthened to accommodate four alternating images—two dragons and two tigers—in galloping poses. Standing by the dragons are, respectively, a bird in flight and a feathered immortal. On the tigers' corners are toads. One of the toads has a bird hovering above, and the other a vegetal motif. The tails of the main zoomorphs slant at a steady angle, adding to the sense of movement. While these thin raised lines lend a primeval quality to the animated beings, they do display a set rhythm and help to establish the basic pattern for other mirrors to come.

22.

重圈神獸鏡
Mirror with Concentric Circles, an Immortal, and Auspicious Animals
Western Han
(206 BC–AD 8),
1st century BC–1st
century AD
Weight: 398 g
Diameter: 13.8 cm
Rim height:
0.5 cm
Overall height
(knob): 1 cm
1995.303

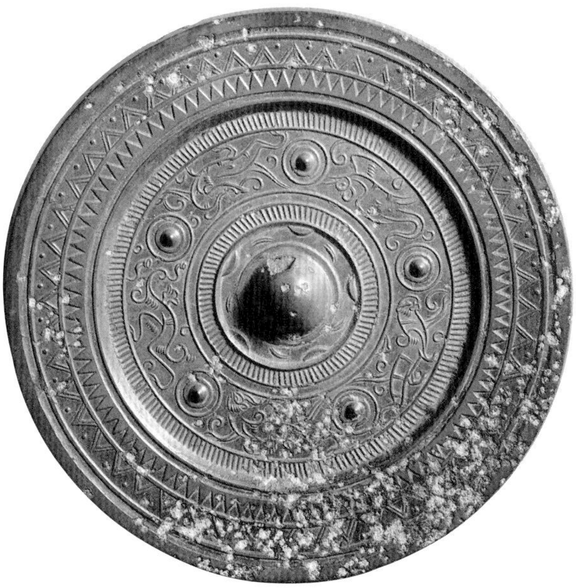

Concentric circles reign supreme in this mirror. Within the inner well, the hemispherical knob on its seating is surrounded by five circular rings, defining an equal number of zones. A hachured band appears twice, one close to the knob and the other at the outer edge. In the principal band, five nipples divide the space to house five different animated beings, all of which are depicted through the use of thin raised threads and lively curvatures. Starting from the dragon (or *qilin*) and moving counterclockwise, we see a bird, a tiger, a unicorn, and a feathered immortal.

In the border, the arrangement is actually of a different order. Two rings of sawtooth design—one with a row of acute angles and narrow bases, the other with a modular unit closer to an equilateral triangle, and also with dual contours and interspersed beads across the row—elicit a diversified appearance. Their density and richness gain a measure of relief when we reach the outer extremity, the rim, with a narrow trough and a plain rounded edge to complete the whole platform. This border treatment, which the Carter mirror clearly shared with, and may have even borrowed from, concurrent mirror types of the time, marked a major departure in the typology and allowed later mirrors produced within the series to catapult to full maturity (see no. 23).

In Henan and Hunan, we find similar, if not identical, mirrors unearthed dating to the Xin era or earlier.[1] Especially in the latter region, a continual sequence can be traced from the Western Han period, through the Xin era, to the Eastern Han. In addition, Guangzhou excavations uncovered one mirror from a late Western Han burial, a version that is in all respects a close relation to the Carter mirror.[2] It would not be far-fetched to settle on a time frame that includes the last decades of Western Han.

1. See Luoyang 1988, pl. 26; Zhou SR 1986b, pl. 44 (Western Han), pls. 47–49 (Xin era), pl. 52 (Eastern Han).

2. For the Guangzhou find, see Guangzhou 1981, vol. 1, fig. 170-(6).

23.

重圈神獸鏡
Mirror with Concentric Circles, an Immortal, and Auspicious Animals

Eastern Han (25–220 AD),
1st century AD
Weight: 719 g
Diameter: 16.3 cm
Rim height: 0.95 cm
Overall height (knob): 1.5 cm
1995.304

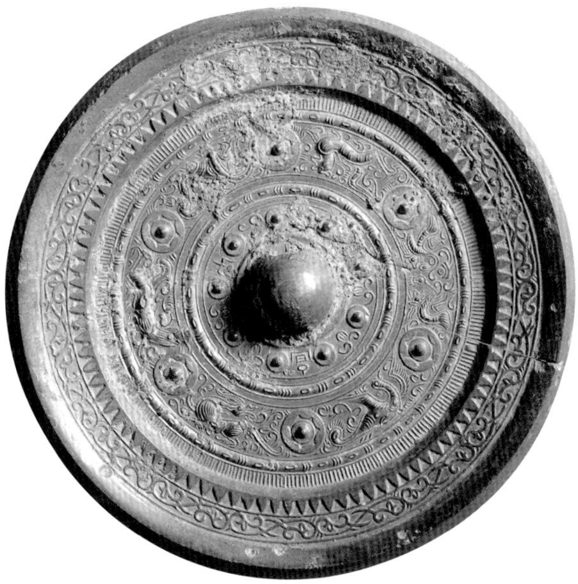

Larger than mirror no. 22, and later in date, this mirror brings a state of refinement to the concentric circular design. Two circular rings of rounded ridges, with the egg-and-dart motif, create a strong accent and divide the inner well into three zones. In the interior zone, a hemispherical knob is encircled with nine nipples and a two-character inscription, each character separated by two unknown symbols. The inscription reads: 宜子[孫] (May your sons [and grandsons] cherish it). The middle zone includes six nipples, which are surrounded by a circle with a sunburst design within. Starting with a pair of immortals, which is perhaps the most noticeable of the motifs, and moving counterclockwise, are a bird, a dragon, a toad, an intertwined tortoise and snake, and a tiger. Four of the set—technically the Red Bird, the Blue Dragon, the Dark Warrior (the tortoise and snake), and the White Tiger—certainly belong to the Four Spirits, which rule over time and space. The toad and the immortals are but additions to the quartet. A narrow ring of hachured lines sets the border for the inner well. The outer zone is composed of a band of sawtooth design, a continuous interlocking vine pattern that vaguely reminds us of zoomorphic origin, and a plain rim raised to a ridge which then descends in a widening sphere. The cross section of that rim would indicate a triangular profile, hence the term *sanjiaoyuan* 三角緣.

Exquisite casting in low relief, sensitive design elements, and free rhythmic interplay as well as contrast and opposition, contribute greatly to the effect of this mirror, which is rich, complex, and elegant. A far simpler mirror, already mentioned, with seven nipples and zoomorphic motifs, carries a date of AD 64; this should help to set an upper limit for this Carter version, and place it in the third or fourth quarter of the first century.[1]

1. See Umehara 1943, pl. 4.

24.

蟠螭紋博局式鏡

TLV Mirror with Serpentine Interlaces

Western Han (206 BC–AD 8), late 2nd–1st century BC
Weight: 302.5 g
Diameter: 14 cm
Rim height: 0.5 cm
Overall height (knob): 1 cm
1995.283

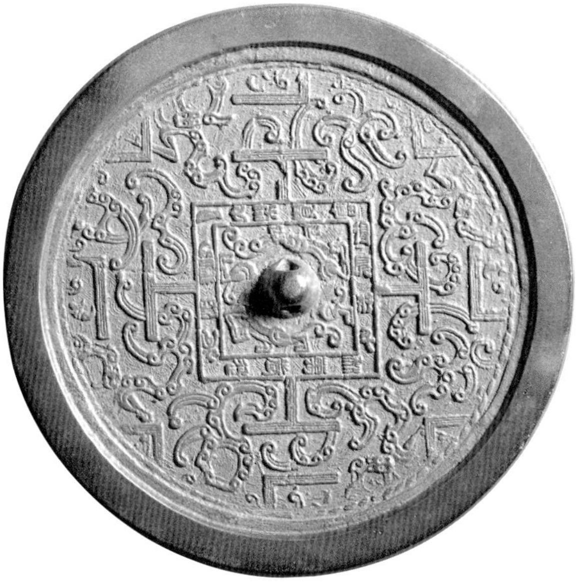

The so-called TLV mirror derives its imagery from the *liubo* game board, thus the appearance of the central square and the T's, L's, and V's beyond it.[1] In this Carter mirror, all these features are present. There is also an orchestration of the *panchi*, or serpentine interlace, that covers every area so defined and circumscribed.

Besides the plain TLV mirrors, of which only a few survive,[2] this was the earliest version of the TLV mirror to appear as well as to disappear. In its heyday, it enjoyed the patronage of the rich and powerful. For that reason, such a mirror found its way into the tomb of Dou Wan, wife of the Prince Jing of Zhongshan, who was buried in 104 BC.

That particular find in Mancheng, Hebei, also enables us to supply a chronological framework for this type.[3] If its peak was reached in and around 104 BC, then it had to begin its life prior to that date, vanishing without a trace before Western Han ended. With this type, the range is limited. Consequently, the design in Dou Wan's mirror is identical with that of the Carter version. To be sure, the Carter mirror's diameter is smaller by 4 cm. Instead of diminishing in size proportionately for both mirror and decoration, the maker of the Carter version appeared to utilize the same (or a similar) template, but cut a smaller circle through it by shortening the diameter. This has resulted in some loss of detail, but mainly on the periphery. Within the band of the central square, the same inscription as in Dou Wan's appears in a script between the *zhuan* (seal script) and the *li* (clerical script):

大樂富貴，得所好。千秋萬歲，延年益壽。

In great joy and fortune,
May you obtain your heart's desire.
Through thousands of autumns and
myriad years,
May you live a lasting life.

1. For the game of *liubo*, see Xiong 1979, 35–39; for the linkage of TLV mirror with the *liubo* board, see the introductory essay in this volume, n. 11.

2. For a direct transposition of the *liubo* game board to mirror in its plain and unadorned state, see Ma et al. 1998, pl. 57. This piece was excavated in Changsha, Hunan province, and is currently in the collection of Hunan Provincial Museum.

3. The best reproduction of Dou Wan's mirror can be seen in Hebei 1996, color pl. 3, pl. 15.

25.

鎏金博局式鏡

TLV Mirror with Gilt Surface

Western Han (206 BC–AD 8), 2nd century BC
Weight: 511 g
Diameter: 14.2 cm
Rim height: 0.6 cm
Overall height (knob): 1.2 cm
1995.294

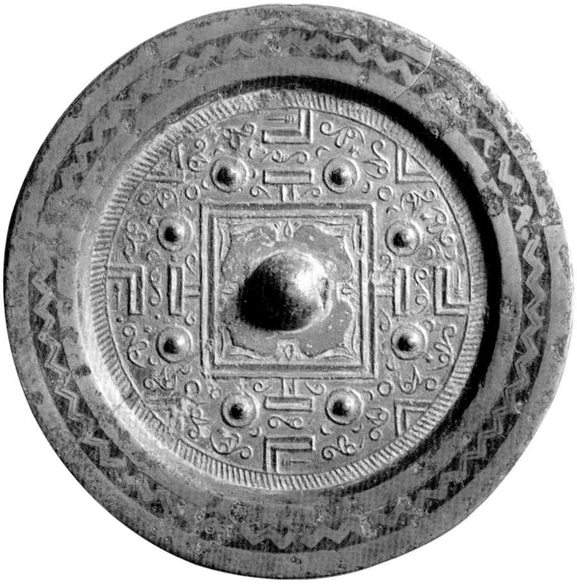

Gilt mirrors of the TLV type are already known from the early Han period;[1] thus, this mirror should not be deemed a novelty. Here the mirror displays the same T's, L's, and V's, but the decorations among them are neither serpentine interlace nor recognizable figures. Rather, sets of curlicues, sometimes simple and other times complex, vaguely suggest some form of living organism, but never supply enough attributes to allow us to identify them as any images other than what they are. Abstraction is the key to this approach.

The impulse to identify the design as a dissolution of recognizable images apparently has been fairly strong among scholarly circles. That is misleading and also misdirected. In fact, during the last phase of Western Han, mirrors displaying similar traits have already made their appearance as a separate branch of the TLV mirror.[2] While not as successful as other branches, this type managed to survive for no less than two centuries. Consequently, sites in Guangzhou yielded examples throughout the whole of the Eastern Han era.[3]

1. For examples of the Western Han era, see Ma et al. 1998, pls. 57, 58.

2. See Luoyang 1998, pls. 23, 24.

3. See Guangzhou 1981, vol. 1, figs. 210-(4), 276-(5).

26.

四神博局式鏡

TLV Mirror with Four Spirits and Companions

Western Han (206 BC–AD 8), 1st century BC–1st century AD
Weight: 1,099 g
Diameter: 19 cm
Rim height: 0.7 cm
Overall height (knob): 1.2 cm
1995.301

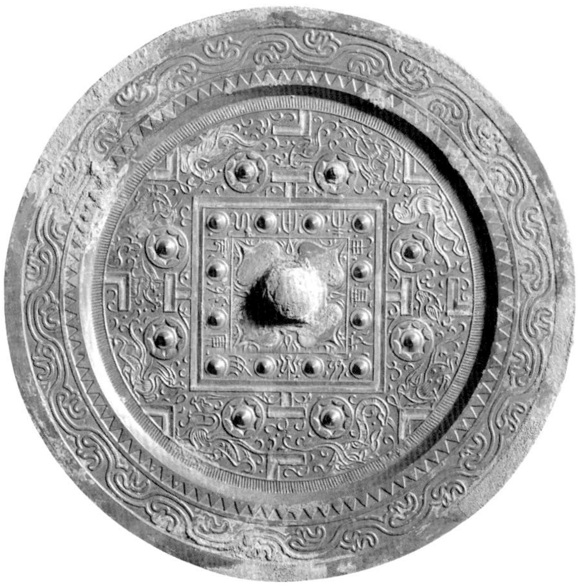

This mirror, along with the next four (nos. 27–30), represents the main branch of the TLV typology. It is undoubtedly the most popular of the TLV mirrors and can be found across China as well as beyond her boundaries. It is also far more complex in terms of overall structure and evolutionary history than those previously discussed. While the game board and TLV disposition remains the same, this branch began with the installation of the Four Spirits in the quadrants beyond the central square and then evolved into the classical configuration of the octet (eight figures).

The reason for this increase from quartet to octet can be easily summarized. Here is a case where design fundamentals prevail over iconography. First, let us assume that the TLV mirror of this branch began with only the Four Spirits. In such a mirror, however, the TL axes were found to be less hospitable when the figures were to be placed between their horizontal bars.[1] Those spaces provided far less liberal quarters than the two side pockets to the left and to the right of the TL axes. When so placed, the Four Spirits would experience a major constriction in their middle, whereas their heads and hindquarters could indeed move about quite freely.[2] Second, in such an arrangement the two side pockets would contain unused space. It is this unused space which must have inspired a strong, additive desire in the mirrorsmith. As a result, he decided on nipples with their bases to fill in the void, and brought in additional figures such as feathered immortals or other creatures. Third, once these figures found their way into the TLV configuration, the next logical step would have been to reduce the scale of the Four Spirits so that they could be accommodated on just one side of the TL axis, allowing the rest, now enlarged in scale, to occupy the other side.[3] In this way, the classic scheme of the octet was born.

The significance of this Carter mirror lies in the fact that its maker covered the Four Spirits with a silvery coating. Obviously he was still keenly aware of the symbolic dimension of the Four Spirits and purposefully accentuated them, even though the reality of their diminishing physical standing was quite apparent. This is in stark contrast to the makers of later mirrors, who, by the last decades of the Western Han period, had already embraced fully the octet arrangement and saw no need for further complication. This is one of our reasons for dating the Carter mirror a step or two ahead of the rest.

The inscription is limited to the area within the central square and among the smaller nipples. It is the zodiac series: 子丑寅卯辰巳午未申酉戌亥. Later TLV mirrors tended to add a second ring of inscription between the inner well and the border.

1. For TLV mirrors with only Four Spirits, see Kong 1992, 280–82. These mirrors are fairly early in date, as witnessed by their plain border, and by the nature of the inscription and its placement within the square band.

2. Even though mirrors with Four Spirits are mostly found in early Eastern Han burials—and in one case could be dated as late as AD 136—the relative chronology of this type definitely prefigures the eight-figure arrangement in TLV typology; see Guo 1996, pl. 62. One of the reasons for this type to continue as late as AD 136 is that a favored mirror type was often replicated over and over again. Dating a mirror actually involves two aspects: first, the duration of the type which it represents, and second, the date of the specific mirror in question.

3. Unlike the Four Spirits, these companions were not as constant, and their linkage to the individual spirits tended to be arbitrary. Consequently, the feathered immortal may accompany the Blue Dragon in one mirror, and the Dark Warrior in another.

27.

多乳博局式鏡

TLV Mirror with Multiple Nipples

Xin (9–24)
Weight: 756 g
Diameter: 16.3 cm
Rim height: 0.6 cm
Overall height (knob): 1.2 cm
1995.295

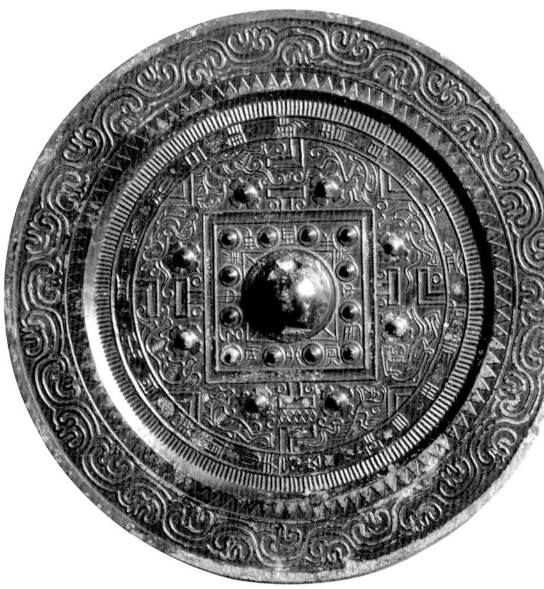

Two sets of inscriptions appear on this TLV mirror. In addition to the zodiac series in the central square, the mirrorsmith had this to say to the prospective owner:

新有善銅出丹陽，
和以銀錫青且明。
左龍右虎主四澎(旁)
八子。

During this Xin dynasty, good copper is being mined at Danyang.
Mixed with silver and tin, the alloy is clear and bright.
The dragon at left and tiger at right:
They govern the four directions.
Your eight sons [and nine grandsons will take the center].[1]

Beyond pinpointing a site for mining copper and the invocation of two of the most powerful of the Four Spirits, the dragon and the tiger, the mention of the Xin dynasty (AD 9–24) places the mirror in the interim period between the Western and Eastern Han eras, when Wang Mang usurped the throne briefly.

Evidence is plentiful that the TLV mirror had evolved to a degree of complexity during the later phase of Western Han. Under the Xin, the tendency was to continue the known trend. Consequently, it should come as no surprise that this mirror resembles mirror no. 26. To be sure, there are some additions and also inadvertent distortions. With respect to the former, we have already mentioned the two sets of inscriptions on this mirror's back: the zodiac series within the central square band, and the second inscription in a narrow ribbon around the central configuration. Relative to the latter, note the dislocation of the Four Spirits. Clockwise, the correct arrangement should be: the Blue Dragon (East), the Red Bird (South), the White Tiger (West), and the Dark Warrior (North). In this Carter mirror, however, the sequence has become Blue Dragon, Dark Warrior, Red Bird, and White Tiger. Of the four, the most difficult to decipher is the Dark Warrior, the combination of snake and tortoise. The snake has a diminutive form and resembles a worm, and the tortoise is even less tortoise-like. They are not intertwined as before, and instead seem quite timid, especially in comparison to their companion, the raging bull.

Perhaps this is due to the mirrorsmith's humorous touch. But let us also note that the nonchalance that has led to the dislocation of the Four Spirits does not seem to prevail elsewhere. The surface treatment of the mirror back is as rich and as elaborate as one can possibly achieve. It also includes eight additional nipples in the area outside of, but adjacent to, the central square, each with a quatrefoil base that further impinges upon the depicted images. The mirrorsmith did not wish to leave any empty space, and thus even inserted the zoomorphs' tails in between the narrow spaces created between the T's and the L's. In fact, in the space governed by the Dark Warrior, he utilized a series of lozenges to fill up the very same area where the diminutive snake turned into an ineffectual worm.

1. As is customary, the mirrorsmith here abbreviated the last line to only two characters of "eight sons." By consulting other known examples, we have filled in the lost words.

28.

多乳博局式鏡
TLV Mirror with Multiple Nipples

Xin (9–24)
Weight: 635 g
Diameter: 16.9 cm
Rim height: 0.5 cm
Overall height (knob): 1.1 cm
1995.298
(Plate 3)

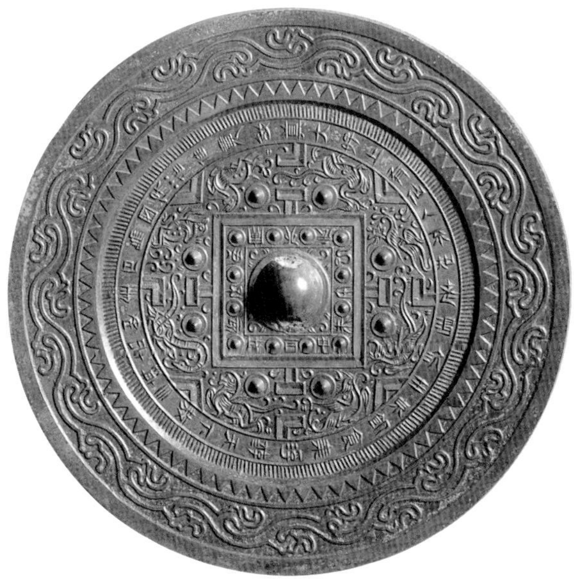

This TLV mirror is similar to the previous examples, but thinner.[1] Consequently, the dual division—the inner well and the border—is diminished in a physical sense. The relief also is shallow but subtle, while the images are made more complex and well organized. Otherwise, the hemispherical knob, the twelve nipples, and the same number of zodiac signs are very much alike in arrangement and in disposition with previous examples (nos. 26, 27), with minor touches and accents. Outside the central square, eight additional nipples once again are present; seated on a circular base, they are not as intrusive as they could be. The T's, L's, and V's are in their rightful place and combinations, indicating the spaces for the four pairs of animals. Here the Blue Dragon has as its companion a goose; the Red Bird gains a twin of equal flamboyance; the White Tiger claims a docile and plump deer; and the Dark Warrior, or the combined tortoise and snake, is accompanied by a feathered immortal.

Despite the rich and elaborate scene, the mirror exhibits an undertone of calm, imperturbable in its equilibrium. With respect to inscriptions, the zodiac symbols remain within the inner square. The inscription in the outer band may differ from the previous examples in the Carter collection, yet it is prevalent among other Han period mirrors of the same category. It reads:

作佳鏡哉真大好，上有仙人不知老。
渴飲玉泉飢食棗，浮遊天下傲四海。
壽如金石為國保。

A fine mirror it is, my making of it!
I can show the ageless immortals up above.
In thirst, they drink from the jade spring.
When hungry, they partake of jujubes.
They wander over the earth and belittle the Four Oceans.
They live as long as metal and stone,
And are the true protectors of our kingdom.

1. For its dating and linkage to Shaoxing, see the explanatory note under no. 29.

29.

多乳博局式鏡

TLV Mirror with Multiple Nipples

Xin (9–24)
Weight: 549 g
Diameter: 16.6 cm
Rim height: 0.45 cm
Overall height (knob): 1.1 cm
1995.299

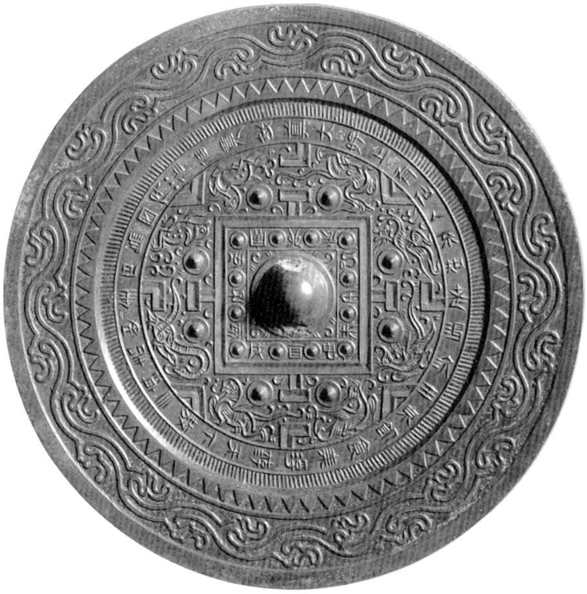

In terms of its dimension, thinness, and rich and calm disposition, this work is nearly identical with mirror no. 28. Even the inscriptions are identical; the only variance comes in the style of calligraphy. While it is possible that the makers of these two mirrors might have worked together within the setting of a private workshop, or in the same region, this craftsman was not without a few unique touches of his own.

Indeed, the most intriguing aspect here is that three of the four principal spirits are now at the left of the T and L axes, the only exception being the Dark Warrior (intertwined snake and tortoise), which remains at the right. Here the spirit's companions are, starting from the Blue Dragon onward: a galloping deer, a feathered immortal, a lively bird, and an upstanding toad. That the arrangement is identical with a purportedly excavated specimen from Shaoxing now in the Shanghai Museum, with a date of AD 15, is important.[1] It signifies a similar chronological framework, most probably Xin, or at the latest, early Eastern Han, and also a geographical linkage. Shaoxing was well known as a major center for mirror production during the Han period, and the fine quality of this mirror testifies to the high level of craftsmanship involved. It is possible that mirror no. 28 could also have been produced at the same place and around the same time.

1. See Chen P 1987, color pl. 38, fig. 38 (rubbing).

30.

方格神獸鏡

Mirror with a Central Square, an Immortal, and Auspicious Animals

Eastern Han (25–220), 2nd century
Weight: 620 g
Diameter: 18.5 cm
Rim height: 0.8 cm
Overall height (knob): 1.7 cm
1995.329

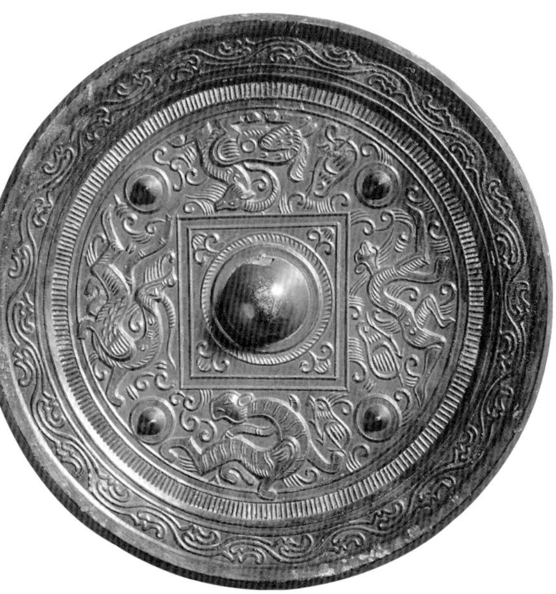

The decline of the TLV mirror during the later phase of Eastern Han allowed new typologies to emerge. A few of the newcomers still betrayed its lasting legacy, as in the prominence of the central square in this Carter mirror. Deprived of the original T's, L's, and V's, the mirrorsmith utilized four pointed nipples to effectively demarcate four spacious compartments. These were used to accommodate a dragon, a tiger, a unicorn, and an immortal, and enabled them to gain in size and dimension. These figures too remind us of the old octet in the TLV mirrors, except that, with only partial adoption, coherence of the symbolism and iconography is lacking.

Comparable examples can be found in Shaoxing's excavated mirrors or those said to be from that region.[1] This mirror comes from a largely Wu-Yue regional heritage, and its time frame is set toward the middle or later phase of the Eastern Han period.

1. For excavated examples, see Wang SL 1987, pls. 40–43, 45–49, 51–53.

31.

方格神獸鏡
Mirror with a Central Square, Immortals, and Auspicious Animals

Eastern Han (25–220), 2nd century
Weight: 643 g
Diameter: 18.35 cm
Rim height: 0.7 cm
Overall height (knob): 2 cm
1995.330

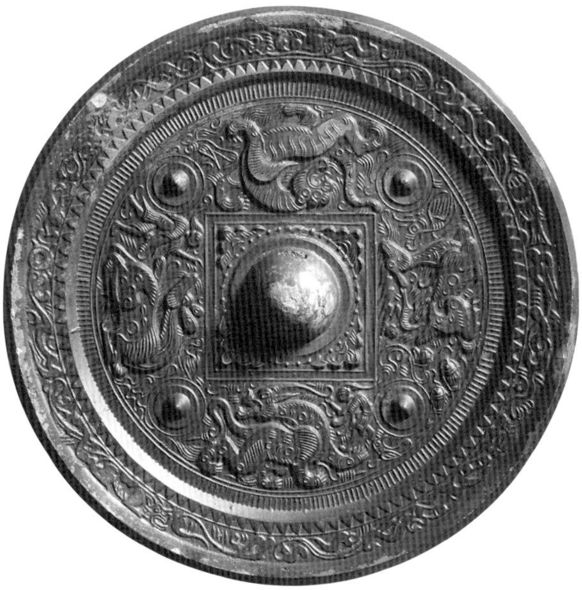

Essentially similar in design and approach to mirror no. 30, this example offers several adjustments. Two deer have taken the place of the immortal and the unicorn. The latter disappeared, whereas the former split into two: one serves as the companion of the dragon and the other accompanies one of the deer. The assumption remains that this probably was a product of Shaoxing and is of Eastern Han vintage.

More so than in no. 30, this mirrorsmith tended to reduce the organic forms of the animals into broad, flattened components. It is almost as if, in making the model for this mirror, he slapped the flat strips of clay onto its back before working out the details. Another element of surprise lies in the treatment of the border's outer ribbon, where the zoomorphic images are squeezed by the narrow confines into curvilinear abstractions. Thus the apparitions of a tiger, a dragon, a fish, and so on, replace the usual cloud forms. In addition, the rim begins to rise at an angle and then slants downward on the side. This is the so-called *sanjiaoyuan* 三角緣, a feature seen in late Han and post-Han mirrors.

32.

畫像鏡

Mirror with Deities and Kings

Eastern Han (25–220), 2nd century

Weight: 847 g

Diameter: 19.5 cm

Rim height: 1 cm

Overall height (knob): 1.9 cm

1995.331

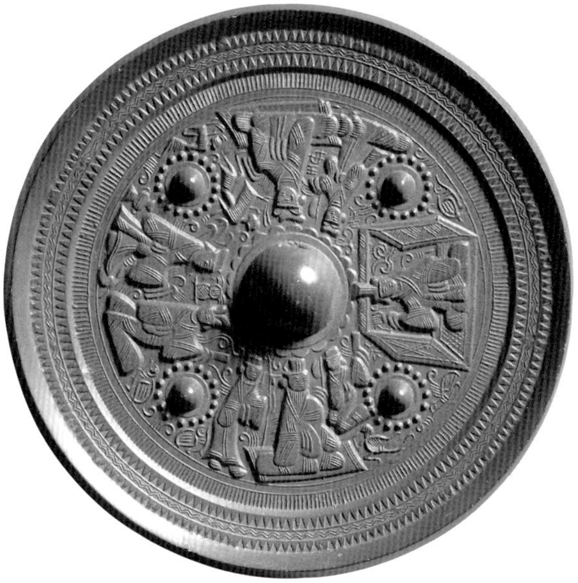

Despite its seemingly intact presence, this mirror endured extensive damage and losses. X-ray inspection reveals that the mirror was broken and then mended. Afterward, it was painted to simulate the effect of *heiqigu*, or black lacquer antique.[1]

The loss of the central square testifies to a major breakthrough in mirror design. Unlike in mirror nos. 30 and 31, the umbilical cord to the TLV mirror was cut and the revolving images on the mirror back gained in size and importance.

On the back of this mirror, the hemispherical knob and four nipples divide the inner well into four sections. With the aid of the inscription, it appears that the Queen Mother of the West 王母 and the Jade Maiden 玉女 are being portrayed together in one section. At the opposite end of the axis, the King Father of the East duly makes his appearance. This leaves the two other sections free for historical kings: Fucha 夫差, the King of Wu, is enclosed within the threefold screens, alone and filled with immense pride at his victory;[2] Goujian 句踐, the defeated King of Yue, has by his side Fan Li 范蠡, his loyal servant and wise counsel. Between them on a scaffold is an object resembling a pineapple, but which actually should be the *dan* 膽 (gall), an organ known for its bitter taste. Goujian, so the story goes, tasted the gall daily and slept on a pile of firewood to underscore the humiliation of his defeat and keep his spirit keen for revenge. Thus the defeated should rise again, just as the victor was to face a reversal in fortune.

1. See the Appendix in this catalogue, under CMA 1995.331.

2. See for instance the mirror narrating the story of Wu Zixu in the Shanghai Museum collection, where the solitary King Wu is also similarly surrounded by a cubicle, his hand raised; Chen P 1987, pl. 51, of Eastern Han.

33.

畫像鏡

Mirror with Deities, Chariot, and the White Tiger

Eastern Han (25–220), late 2nd–3rd century

Weight: 503 g

Diameter: 18.3 cm

Rim height: 0.85 cm

Overall height (knob): 1.7 cm

1995.332

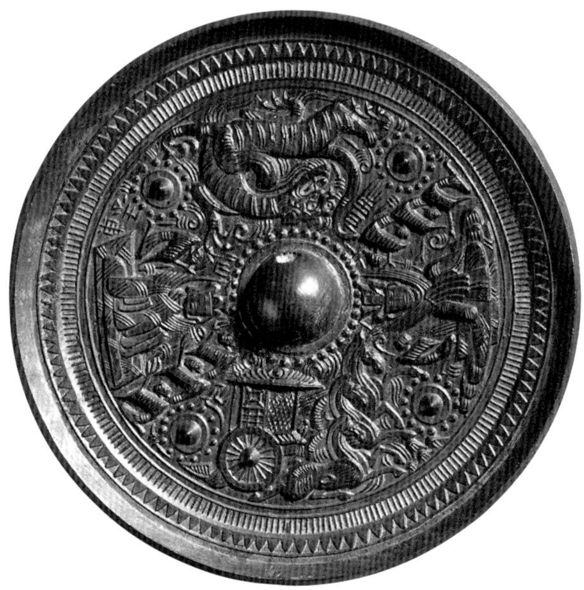

Short inscriptions on this densely populated mirror back identify one deity as the King Father of the East 東王公 and the zoomorph as a White Tiger 白虎. The presence of the former demands his opposite, namely, the Queen Mother of the West. The White Tiger, however, is not accompanied by any of the three remaining Four Spirits. Instead, a chariot drawn by three horses is pictured at its opposite end. The hemisphere of the knob and its base is surrounded by a circle of small nipples, which is echoed in the four large nipples farther away, also similarly surrounded. The large nipples are retained more as vestigial remains than as functioning dividers.

One cannot but question the wisdom of displaying both the Queen Mother of the West and the White Tiger. As symbolic of the West, they are redundant. This, of course, is not uncommon, for it was equally dubious to portray a Blue Dragon on the same mirror with the King Father of the East, as the former is also the rebus for the East.[1] The chariot moves in the direction of the Queen Mother of the West; scholars have interpreted this as representing King Mu of the Shang dynasty and his fabled visit to this deity.[2]

1. See Wang SL 1987, pl. 30. A number of this type of mirror were unearthed in the vicinity of Shaoxing, some displaying two opposing chariots; see ibid., pls. 26–30.

2. Ibid., "Explanatory Notes," p. 5, under entry 27.

34.

雙夔連弧鏡

Mirror with Linked Arcs and *Kui* Dragons

Eastern Han (25–220), 2nd century

Weight: 465 g

Diameter: 15.3 cm

Rim height: 0.45 cm

Overall height (knob): 1.2 cm

1995.320

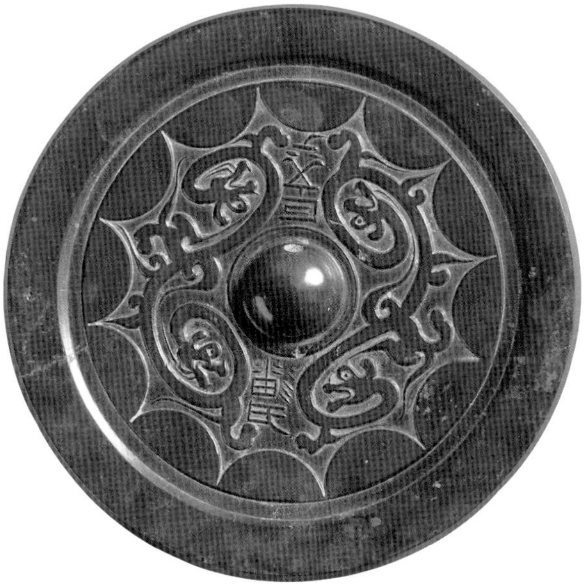

Bearing the patination known as "black lacquer antique," this particular mirror features a simple but effective design, with a central axis occupied by the inscription. The inscription is divided by the central knob into two different sectors, each to be read away from the hemispherical knob. Flanking the axis are two pairs of *kui* dragons in roughly parallel S-curves, at the extremities of which are dragon heads. Surrounding this complex is a series of twelve linked arcs forming a sunburst pattern. A double contour of circular bands divides the inner well from the border, which is also quite plain.

With the craftsman claiming authorship, this mirror bears the inscription 潘氏作文兮 (I, Pan, have produced the design). Of the mirrors made by this master, we know only of this and one other example, a specimen reproduced by Luo Zhenyu in his *Gu jing tu lu* (Illustrated Catalogue of Ancient Mirrors).[1] That one is a TLV type, in which the Four Spirits and their counterparts are replaced by eight birds; they position themselves away from each other across from the T's and L's. It also has a lengthy inscription, beginning with the sentence:

潘氏作鏡世少有，服者男分封候...
宜孫子，樂無巳。

Master Pan's mirror is rare in this world.
He who owns it would attain the rank of marquis . . .
It should be cherished by your descendants.
May your joy be endless.

Is this the same Master Pan who produced the Carter mirror? Could he be of the same clan but of different generations? With only two unprovenanced mirrors at our disposal, we are handicapped by the lack of sufficient evidence to make a firm determination.

The Carter mirror is of a type dating to the Eastern Han period by virtue of archaeological parallels.[2] Regarding the TLV type with eight birds, an example was unearthed in Guangzhou and is also dated to the late Eastern Han.[3] However, since Guangzhou is located in a remote hinterland, one may assume that this type of mirror might have had an earlier start in the central part of China. Another was found in Hengyang, Hunan, in a tomb of the Southern dynasties period.[4] The inscription declared that it was made by the Shangfang Bureau, an official foundry. If so, Master Pan in his practice may have adhered to the design that was of a courtly origin.

1. Published in Luo 1916, *juan* 2, 16b.

2. See Kong 1992, 464, 468. The former was unearthed in a Shaogou site near Luoyang and is dated to the late Eastern Han by virtue of the burial's age. The latter is from Shunhua, Shaanxi province.

3. Kong 1992, 321.

4. Ibid., 322. For an archaeological report, see Hengyang 1986, 1079ff, especially p. 1091, fig. 19.

35.

雙夔連弧鏡

Mirror with Linked Arcs and *Kui* Dragons

Eastern Han (25–220), 2nd century
Weight: 255 g
Diameter: 12.5 cm
Rim height: 0.35 cm
Overall height (knob): 1.3 cm
1995.286

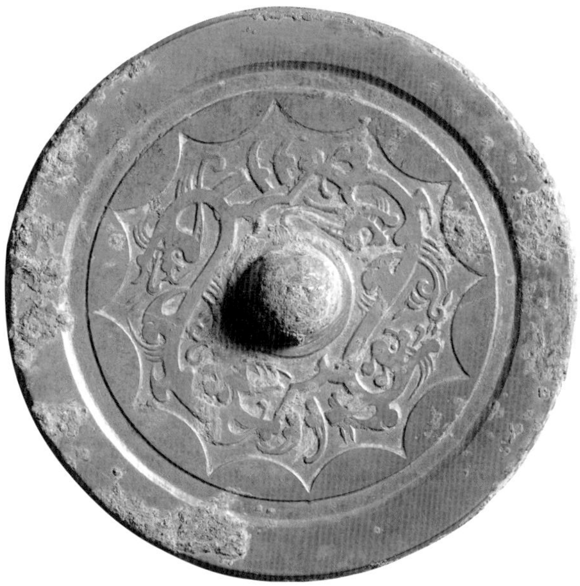

The silvery appearance of this mirror is in sharp contrast to Master Pan's black luster (see no. 34). Otherwise, the design is similar if not entirely identical, testifying that neither Master Pan nor anyone else had a monopoly on this particular type. Perhaps the arcs here do not arch as high. But the absence of inscription allows the mirrorsmith to integrate far better two parallel S-curves, thereby creating a sensation of undulating revolution within the sunburst image.

36.

四葉連弧鏡

Mirror with Quatrefoil and Linked Arcs

Eastern Han (25–220)
Weight: 235 g
Diameter: 12.2 cm
Rim height: 0.3 cm
Overall height (knob): 1.1 cm
1995.319

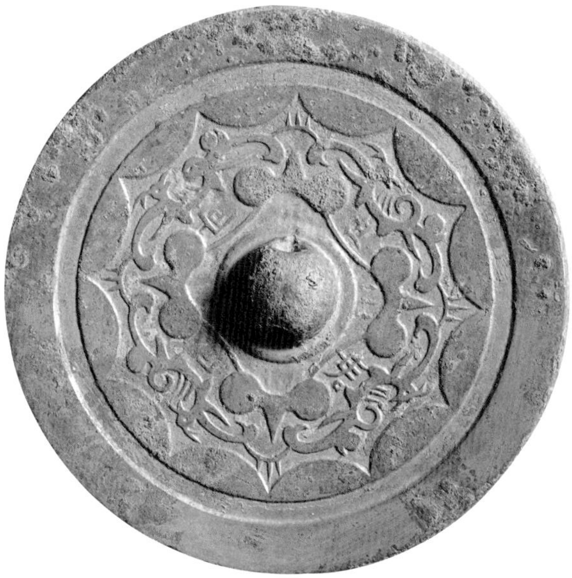

Though corrosion has taken its toll, we cannot ignore this mirror, which represents a transitional stage in the evolution of a highly elaborate and, for want of a better term, "psychedelic" approach to mirror design (see no. 37).

Herein lies the significance of this particular work, which shares features with mirror nos. 34 and 35, as well as no. 37. It is obviously a step beyond the former mirrors, though still a long distance from the latter. While it has the hemispherical knob, the quatrefoil, the sunburst image, and the plain border, nonetheless its quatrefoil motif clearly has expanded outward while still joined by curves that begin and end at the apex of each of the twin petals. These curves, when seen together, form a lozenge that holds the knob at its very center. The former element of the *kui* dragons has now been pushed to the edge, losing their central position. They, too, have been transformed. Looking from a centripetal point of view, each has turned into something resembling a quadruped with a bird head, with indecipherable torso and limbs, as well as appendages.

In evolutionary terms, the critical design step consists in turning the quatrefoil into a core image that can expand and contract rhythmically. With that step taken, the mirror type is ready to move on to a further stage, one that can cause the mirror design to throb and pulsate in ways that permeate the entire platform.[1] This was a major triumph, and a new branch of mirror design was born of this seemingly humble origin.

The inscription is a simple one, conveying well wishes from the mirrorsmith to the owner: 君宜官位 (You deserve an official post).

1. See mirror no. 37, which is a branch in the typological development from patterns shown in this mirror. Far better than this example are those mirrors that indeed throb and pulsate with organic forces.

37.

四葉瑞獸對鳳鏡

Mirror with Quatrefoil, Paired Phoenixes, and Auspicious Animals

Three Kingdoms (220–280) to Western Jin (265–316)
Weight: 694 g
Diameter: 18.6 cm
Rim height: 0.4 cm
Overall height (knob): 1.2 cm
1995.317

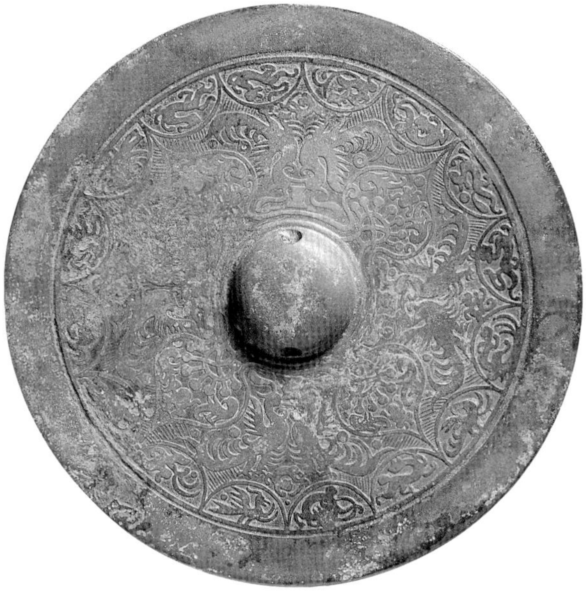

This mirror's knob is broad in shape and flattened at the top—a visible change from the earlier hemispherical form. The border is quite plain. However, between the core and the periphery, the decorative pattern is rich and elaborate, mostly done in one simple level of relief. From the knob, the quatrefoil has been extended; the tips of the four petals insert themselves in four points, where two arcs join at the enclosing circle. Thus, the division into quadrants is clear.

Within each petal, close to the stem, is a curious bulb-shaped motif with a diagonal cross at the lower portion and seven projections fanning out from its lower edge, each with a round end. The rest of the petal area is given first of all to a small toad in a frontal pose at the center and over a circle that may represent the moon. At two sides are found birds, whose bodies face opposite directions but whose heads turn to each other. Outside the petals and between them are four pairs of proud phoenixes, with high crests and full plumage, facing each other in profile with beaks and chests touching in a tight bond. Underneath their claws, a small mammal, perhaps a monkey, curls its body into a circle.

A series of sixteen arcs line the enclosing circle and provide space for additional zoomorphs. The sixteen creatures are not always recognizable. The sequence, in the segments that are legible, consists of the following: phoenix, dragon, tiger, long-tailed animal, monkey, fox, long-tailed animal, rooster, dog, monkey, phoenix, bear. Could these be the combination of Four Spirits plus the twelve zodiacs? The presence of the bear would seem to negate this easy solution, as that animal belongs to neither the former nor the latter.

In spite of corrosion, this mirror is fairly clear in its design and is the only example in the Carter collection of a type that prevailed from the mid Eastern Han to the Six Dynasties period. In China, it is often named with a prefix of *bianxing*, meaning "mutated." Mutation, of course, is not exactly a fitting term, as it implies a degree of deformation from a normal pattern. But this and other similar specimens are just as "normal" in that they form a distinct category. Within the larger type are further subtypes, which depend on the treatment of the diverse parts. The present mirror may be dated to the Jin period or shortly before, i.e., the Three Kingdoms period, while the closest parallel may be found in a mirror unearthed in a Western Jin tomb in Xin'gan, Jiangxi.[1] The flattened knob, the alternating petals, the paired phoenixes, and the sixteen arcs lining the circle are common features. Even the motif of the bulb shape, with its seven projections and the pair of birds framing a small toad, is also repeated here. The only difference between the two is that this archaeological find's border is not plain, but instead completely filled with zoomorphs along both the crest and trough of a long, undulating vine. That being said, it is also useful to remember that the plain border encircling this general typology can be seen in mirrors dating to a time prior to AD 249, and even into the first quarter of the fourth century.[2] Those attentive to archaeological finds may also consult the mirror found in Lucheng, Yidu,[3] or even the mirror from the Wu state during the Three Kingdoms period to verify the dating.[4]

1. Kong 1992, 402. See also Xin'gan 1983, 1122–24, fig. 4; this complex mirror adds one more layer of the continuous arcs, with one zoomorph within each enclosure. In terms of pattern evolution, it appears that the Carter mirror is earlier than the Jiangxi counterpart, which is dated to the Western Jin period.

2. See Kong 1992, 394, 396.

3. See Yichang/Yidu 1988, 910–17, pl. 6-(3).

4. Zhenjiang 1984, 536.

38.

半圓方枚環繞式神獸鏡
Mirror with Deities and Animals Surrounded by Rings of Squares and Semicircles

Eastern Han
(25–220),
late 2nd century
Weight: 509 g
Diameter: 15 cm
Rim height: 0.45 cm
Overall height
(knob): 1.3 cm
1995.333

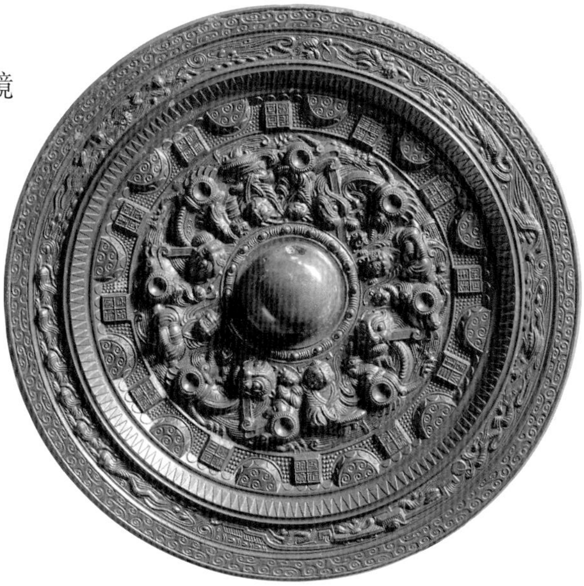

As befits its name, this Carter mirror is noted for the rings of twelve squares and an equal number of semicircles at the outer registers within the inner well or compartment. In terms of typology, it obviously took as the point of departure those mirrors with linked arcs, then separated the arcs and added squares between them.

This particular feature reduces the capacity of the mirror's inner well to accommodate appropriate themes and subjects. To compensate for the lost space, the mirrorsmith first of all put extraordinary effort into his detailwork, and then aimed for a presentation of higher relief than ever before. Both contribute to the unprecedented density of design.

The generously sized knob is surrounded by a railing of eggs-and-darts, which serves as the core of a series of concentric circles that are the basic pattern of the mirror. Immediately afterward, the principal circular band appears. Four *pixie*, each holding a long scroll in its jaws, divide the band into four quadrants.[1] Each *pixie* also supports on its back a triad of figures. In one instance, the central figure—a player of a musical instrument, the *qin*—is undoubtedly BoYa. If so, the figure at his left must be Zhong Ziqi, who tilts his head listening in rapture. A third figure looks away from the scene. He may be the exact opposite of Zhong, and his inability to appreciate the lofty music eventually compelled BoYa to stop playing his music altogether. In another quadrant, the main figure is seen wearing a *sheng* 勝, a headgear shaped like a long, inverted rectangular "U" with two short arms pointing downward; she is the Queen Mother of the West. Accompanying the Queen Mother is a bird to her left and an official of her celestial court to her right. Between the BoYa and the Queen Mother quadrants is the King Father of the East, whose headgear has three prongs, as in Han reliefs.[2] The only remaining figure, to the right of BoYa, is probably Huangdi, the Yellow Emperor, whose headgear is more intricate and crowned with a reclining "C." His retinue includes two strange zoomorphs. We learn from a number of inscriptions that "BoYa plays music, and Huangdi vanquishes the evil"[3] to affirm the pairing, even though the two were obviously unequal in their stations of life.

Given that all the images are deliberately ordered, one cannot but be puzzled by the recurrence of small circular rings with concave interiors—two in every quadrant. Upon close examination, it is clear that the right ring is set at the shoulder portion of the *pixie*, while the left one is at the rear where the thighbone joins the hip. If this is so, then these are part of the anatomy of these auspicious animals. One wonders if the inset of semiprecious gems into the concavities would have evoked the appearance of bulging muscles even more.

The density and high relief of the pictorial portion place this mirror (and its kin) in a class by itself. The surrounding rings of squares and semicircles also add complexity and precision to the images. The squares are attached to the inner register of the circular ribbon, whereas the semicircles are attached to the outer one. The semicircles contain whorl patterns. The squares, however, resemble seals in that each of them is subdivided into four equal parts, each part filled with a written character and cast in perfect clarity. The full inscription reads:

吾作明鏡，幽湅三岡。配仙世袁，統德序道，
敬奉臣(賢)良。同(銅)刻無極。
百牙作樂，眾祥主陽。世照光明，
壽如金石，位至三公，其師命長。

I made this bright mirror,
By refining three metals.
It is fitting for the immortals, distant from the worldlings,
And for those who are in command of virtues and the Way.
Here too we will find respect for the wise and the able.
The bronze can be engraved in infinite ways.
Here is BoYa playing the music.
A multitude of fortunes underscore the yang *force.*
The mirror shines brightly on the world.
May you live long as the metal and stone.
May you reach the height of dukedom.
Even this master be given to live a long life.

Despite the plenitude of images already present, the craftsman refused to relent. Even at the mirror's border, he decided to include a narrow pictorial band to allow further thematic exploration. Indeed, in this narrow space, action and dynamism reign supreme. Central here is the onrushing of six winged dragons, with a stretched chariot behind them. The three compartments in that celestial chariot display a charioteer, wearing short sleeves and stretching his arms forward to hold the reins; a lordly figure seated in frontal pose, with knees parted wide; and a kneeling figure who pays obeisance to the lord. Before this chariot, traces of clouds lead us to two riders, one on a dragon and another on a tiger, the latter's garments trailing and fluttering in the wind. Behind the chariot, a male deity holds a disc. A giant tortoise and a pair of elegant birds with long necks and spreading wings follow him. Then comes a strange apparition: that of a man seemingly without his lower torso and displaying a menacing countenance, animal-like in its intensity. He is followed by a female figure, who holds up another disc. Between them, the trailing clouds flow rapidly by, while the figures are always in a hurry to force their way forward. The momentum is unstoppable.

Again, we can observe the presence of mythic figures and spiritual forces. Those who hold the circular discs cannot be anyone but Fuxi and Nuwa, the legendary emperor and his consort. The Four Spirits are duly present: the Blue Dragon, a pair of Red Birds, the White Tiger, and the Giant Tortoise of the Dark Warrior. If we could hazard a guess that the lordly figure in the central location of the chariot is Huangdi, the Yellow Emperor, then the only still-unidentified figure, the one with the menacing countenance, may well be Zhiyou, who was evil incarnate and destined to be vanquished. If that should be the case, then the themes of the central and peripheral portions do intersect, and mutually reinforce each other.

Of those mirrors that use the device of a revolving series of squares and semicircles around the periphery of the inner well, the earliest dated example (AD 105) is found in the collection of the Goto Museum, Japan.[4] In view of its fully developed state, one can sense that the typology would have evolved even earlier. Insofar as excavated materials are concerned, this type can be found in archaeological sites of late Han, in the Wu state during the Three Kingdoms period, and during the Jin period.[5] For the Carter mirror, we propose a date of the second half of the second century—the principal reason being its close resemblance in iconography to two mirrors in the Shanghai Museum collection, respectively dated 167 and 187.[6]

1. China's mythic tradition encompasses a number of fabulous animals. *Qilin* is the best known, as its presence suggests universal peace. Its close kin is *pixie*, whose name suggests "warding off evil influence." A *pixie* serves the function of a guardian, and is paired with *tianlu* (heavenly munificence) in front of tombs of eminent personages during the Han and the Northern and Southern dynasties. During later phases, it is quite clear that these two animals' identities became less and less distinct until they were practically interchangeable. See Chu 1935, 183–99, on this issue. Citing literary sources to support his view, Chu makes quite clear that *qilin* is known either with a single horn or without horn altogether. It has hooves rather than claws, and its tail resembles that of an ox. *Pixie*, on the other hand, has two horns and claws rather than hoofs. Its companion, *tianlu*, has only one horn, with a torso said to resemble a deer.

2. For a different interpretation regarding the thematic substance of this and similar mirrors, see Ma 1962, 52–55. One can also consult existing mirrors for determining such symbolic attributes. See Higuchi 1979, vol. 1, 177, fig. 79, where the inscription identifies the deity portrayed as the King Father of the East. He wears a crown with three prongs.

3. For an example, see Chen P 1987, pl. 56, with inscription cited in the text portion, p. 34.

4. For this Goto example, see Kong 1992, 410.

5. For dated Jin pieces, see ibid., 422, 424. The former mirror is dated AD 282, the latter AD 342.

6. For these two mirrors in Shanghai Museum, see ibid., 411, 413.

39.

半圓方枚環繞式神獸鏡
Mirror with Deities and Animals Surrounded by Rings of Squares and Semicircles
Eastern Han (25–220), late 2nd–early 3rd century
Weight: 270 g
Diameter: 11.8 cm
Rim height: 0.35 cm
Overall height (knob): 1.2 cm
1995.326

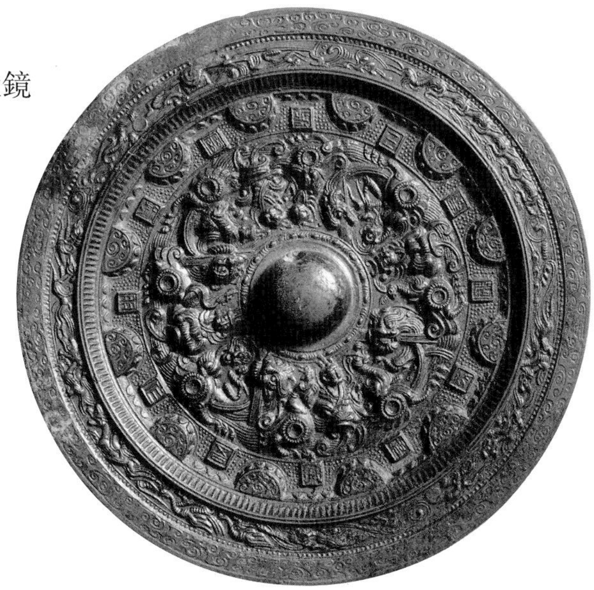

This mirror is essentially identical to mirror no. 38. In the principal band, the four *pixie* still hold scrolls in their jaws, and swing their tails back to hold the triad of figures—among whom, BoYa is instantly noticeable as he plays the *qin*. Of the remaining three figures, one wears headgear that supports a reclining "C"; by our reasoning, he should be the Yellow Emperor. The two other figures all have three-prong headgear. Once again, by referring to mirror no. 38, we can deduce that the King Father of the East is immediately adjacent to the Yellow Emperor, whereas the Queen Mother of the West is closer in position to BoYa. This means that she was erroneously given the male headgear, instead of the usual *sheng*. All the attending figures to these main icons are generally in order.

Also similar to mirror no. 38, the pictorial ribbon at the border contains the heavenly chariot pulled by six dragons. In front of the dragons are three figures; the male deity with the disc must be Fuxi and the two others ride on a dragon and a tiger. Behind the chariot is Nuwa holding the moon. A pair of birds and four ferocious animals in gallop further accentuate the movement.

The inscription utilizes the ring of squares, each of which contains one word. It is but an incomplete rendition of mirror no. 38's inscription:

吾作明鏡，幽湅三商。
雕刻極無（無極）。[1]

I made this bright mirror
By refining the three metals.
Then I sculpted infinite forms . . .

It is clear that this mirror derived its basic structure from a predecessor like mirror no. 38. It lacks the latter's precision in casting and, in presenting the iconographic program, began to reflect some measures of corruption. These signs point to a later date than the previous example, perhaps the last quarter of the second century to the first quarter of the third century.

1. The last two characters must be reversed in order to make sense.

40.

半圓方枚環繞式瑞獸鏡

Mirror with Auspicious Animals Surrounded by Rings of Squares and Semicircles

Eastern Han (25–220), late 2nd century
Weight: 242 g
Diameter: 11.8 cm
Rim height: 0.4 cm
Overall height (knob): 0.8 cm
1995.324

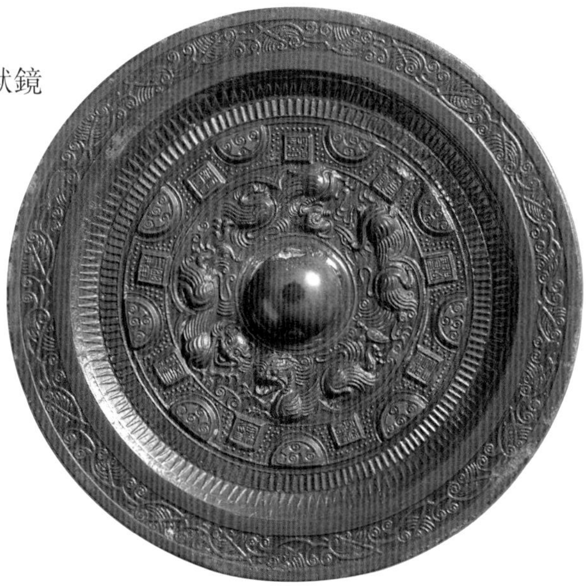

In this mirror the mythic figures and deities disappeared, leaving only two pairs of auspicious animals facing each other. Three are identifiable as *pixie*, having broad faces with whiskers and manes, and powerfully undulating bodies with strong shoulders and hindquarters, sharp talons, and fluffy tails. The lone exception has a narrow dragon-like snout, although it shows little difference in bodily proportions. Deciphering the inscription proves enormously difficult, as the mirrorsmith abbreviated the strokes in the characters and made them almost illegible. The final result reads as follows:

吾作明鏡，幽湅三商。伯牙作樂，
眾神見容。宜大富兮，有福安服。
曾年益壽，子孫番昌，其師命長。

I made this bright mirror
By refining the three metals.
Bo Ya is making music;
The multitude of gods display their forms.
[The mirror] is suitable
For a man with great wealth and good fortune.
He [who has it] will add years to life.
His descendants will also prosper.
Even this master will enjoy longevity.

The closest mirror in terms of imagery, with only four squares and none of the semicircles, is dated by inscription to AD 189.[1] A late second century dating therefore is reasonable for this mirror.

1. See Umehara 1943, pl. 15a.

41.

半圓方枚環繞式三段神人鏡

Mirror Featuring Deities and Kings in Three Sections Surrounded by Rings of Squares and Semicircles

Eastern Han (25–220),
late 2nd–early 3rd century
Weight: 280.5 g
Diameter: 13.9 cm
Rim height: 0.4 cm
Overall height (knob): 1 cm
1995.328

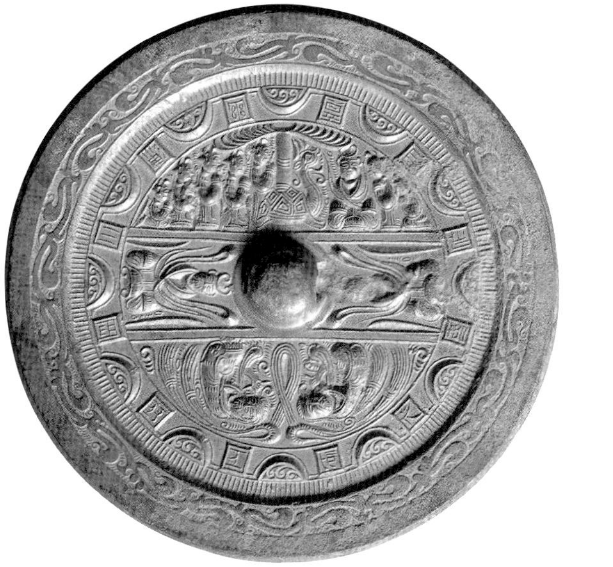

Of mirrors with revolving squares and semicircles, this type is unique in demanding the viewer to see the scenes from differing angles. Very likely, this was in response to the fact that such shapes take up space and reduce the opportunity for iconographic involvement. To use the available space, the maker of this mirror decided on two useful steps. The first was to enlarge the inner well at the expense of the border, an area where the decoration consists of a continuous linkage of curvilinear motifs that resemble dragons only in a far-fetched way. The second was to divide the enlarged interior space into three zones, so as to accommodate the iconic or narrative presentation. Therefore, the central zone, which is further divided into two halves, is to be viewed from the periphery and back to the center. The two crescent-shaped narrative panels have their own respective registers and vertical axes. They are seen from the knob to the periphery. In sum, there are four viewing angles in one single mirror.

The central zone houses the icons of two deities, who are seated with scarves flowing upward toward the knob. A male bearded deity is seen with a crown topped by a reclining "C." Unfortunately, the facial attributes of the other are beyond recognition, though the slender proportions may indicate a female deity. It would not be entirely illogical to assume that they are the divine pair of the King Father of the East and the Queen Mother of the West, even though the former wears a headgear normally reserved for the legendary emperor Huangdi, or for mythic emperors in general.

Of the crescent-shaped panels, one features a pair of kneeling figures surrounded by, if we were to hazard a guess, thick curtains with a divider in the middle. In the other panel, a central pillar supporting a parasol rests on the back of a tortoise. At its right sits a larger figure, wearing a crown with a reclining "C" to indicate kingship. Behind him are two maidens. At the left is a row of five smaller figures, the tallest at the head of the procession and the shortest at the end. The latter is obviously compelled to kneel as the arc of the crescent comes to an end, leaving little space for any standing figures.

The theme of this mirror, apart from the icons of the two deities, remains unknown. To seek an answer, we first consulted the inscription, which proved not at all helpful. Contained in the ring of squares, the inscription is not only brief, but also incomplete:[1]

吾作明鏡，幽湅三岡。巧工刻之。

I made this bright mirror,
By refining the three metals.
With my fine skill,
I engraved [the design].

This forced us to look elsewhere for a solution. In this context, a Sichuan mirror with a similar layout supplied a definite clue. In a longer and substantive inscription, it provided us with the following thematic material:

余造明[鏡]，□□能容。翠羽秘蓋，
靈鵝台□。雕刻神聖，西母東王。
堯帝賜舜二女，天下泰平...[2]

I made the bright mirror,
. . . To portray . . .
Precious parasol of kingfisher feathers,
And the Terrace of the Divine Geese.
I also shaped the deities and sages,
Such as [Queen] Mother of the West, King [Father] of the East,
The Emperor Yao betrothed two daughters to Shun.
Peace reigns on earth . . .

This essentially confirms the pictorial scheme of the Carter version. The pair of King Father and Queen Mother resides in the central zones, their heads pointing to the knob. The crescent panel featuring the precious parasol is a presentation in the rite of marriage. Yao is portrayed at the right, with his two daughters, Ehuang 娥皇 and Nüying 女英. Leading the four figures at left is Shun, the rest being the sons of Yao.[3]

What then is the content of the opposing crescent? Since this panel shows only two kneeling figures, it may very well be the fabled, ancient ritual of *chanrang* 禪讓. This took place sometime after the marriage, when Yao yielded his throne not to his offspring, but rather to Shun, who was sagacious and wise, and therefore more deserving of the kingship.

It is apparent, august theme aside, that this mirror type enjoyed only a limited circulation. Perhaps the multiplicity of viewing angles was deemed confusing and failed to attract a wide clientele. Besides the Sichuan example, so far the rest of the excavated examples come from Xi'an and Qianxian, all within Shaanxi province.[4]

1. See Guo 1996, pl. 55. It contains a fuller inscription, from which the maker of the present mirror decided to select only the first line.

2. See He Z 1991, 1–8, especially p. 6, fig. 17 and pl. 1-(3).

3. Yao is said to have ten sons. Due to space limitations, only four are featured here.

4. See Shaanxi 1959, pls. 75–77. The last mirror (pl. 77) was excavated in Qianxian, also in Shaanxi.

42.

龍虎對峙鏡

Mirror with Two Dragons and a Tiger

Three Kingdoms (220–280)
Weight: 770 g
Diameter: 14.7 cm
Rim height: 1.1 cm
Overall height (knob): 1.5 cm
1995.322

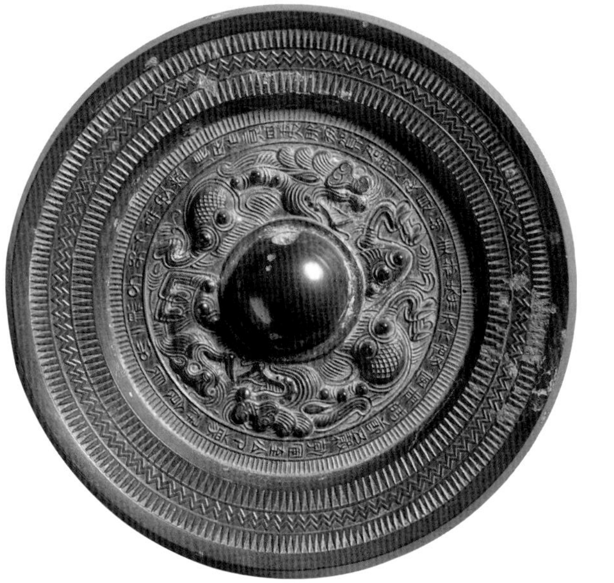

While of modest size, this mirror is notable for its larger hemispherical knob, its thickened outer circle, and its angular incline at the edge. The inner well around the knob is decorated with a tiger and a pair of dragons. The dragons are in procession, one following the other, ending in a dramatic confrontation with the tiger, whose jaws open wide. Both dragons are shown in profile view, with whiskers and triple-braided hair, and a long horn; they also sport a slender middle section, marked by three round bosses, and burgeoning hindquarters with cross-hatching and two round bosses that terminate in a tail. The most confusing figure is that of the tiger. As it turns its head toward us, it appears to have only its front torso, with raised shoulders and forelegs in a crawling posture resting on one side of the knob. But as it extends its hindquarters across the knob, it simultaneously twists its spine in a way that defies belief. A circular ribbon of inscription encloses the pictorial sphere, which in turn is enclosed by another with short hachured strokes. The wide border contains two rows of sawtooth markings, with another row of double-contoured wave patterns between them.

The inscription on the mirror's back identifies a Master Zhang as its maker:

張氏作鏡四夷服，多賀君家人民息。
胡羌殄滅天下復，風雨時節五穀熟。
官至公卿蒙祿食，長保二親子孫力。
樂毋極。

I, Master Zhang, produced this mirror,
When alien tribes in four corners were being subjugated.
Let us praise our Lord, for the people are given respite.
After quelling the Hu and Qiang, he will restore peace on earth.
May you reach the high office and receive your just reward.
May you [gain the power] to protect your parents and descendants from harm.
May your joy be endless.

So it appears that Master Zhang, who had once lived in a time of turmoil, was fortunate to be able to sense the upcoming peace and prosperity.[1] Still, this does not provide a clue as to the specific time period when he was active. Ironically, this verse was by no means Master Zhang's own. Others also have resorted to it—for example, Master Qinggai and Master Long.[2]

Before we attempt to arrive at a chronological framework for the mirror, let us remind ourselves that the name of Master Zhang may not merely signify one individual. There could be several Masters Zhang in existence, all mirrorsmiths within the same clan. Speaking in terms of excavated mirrors, the earliest of those bearing Master Zhang's impression were datable to the mid Eastern Han period,[3] whereas the last dated was cast in AD 273.[4] The visible span is thus approximately 170 years, comprising several generations of craftsmen. We have also learned that one Zhang clansman and mirrormaker, Zhang Yuangong, of late Eastern Han, claimed to be a man from the Wu district (Suzhou).[5] If so, then the rest of the Zhang masters, too, would have been natives of this region.

Even more important, one of the Masters Zhang, or his mirrors, may have traveled to Japan—most likely during the Three Kingdoms period. Inscriptions on a number of excavated mirrors in Japan, of the type known as 三角緣神獸鏡 (Mirror with Deities and Animal with a Slanted Edge at the Rim), announced his authorship.[6]

If Master Zhang cannot help us to date this Carter mirror, then we must resort to stylistic signals for guidance. In this respect, the enlarged knob, the gradually thickening border, and the slanting edge suggest a date from late Eastern Han and before the Western Jin. To narrow the temporal framework still further, we have learned that during the Eastern Han, the dragon was treated in a manner that made him appear to be covered with body plates, and the rounded bosses were even more numerous than those in the Carter mirror.[7] A Three Kingdoms date seems to be a plausible answer to our query.

1. See Liu T 1935, *juan* 15, pls. 47a, b.

2. For the mirror made by Qinggai, see Kong 1992, 480; for Master Long's example, see Feng/Feng 1906, vol. XI. Many of the Shaoxing finds carry the same inscription, and the artists bear surnames other than Zhang; see Umehara 1984a, *passim*.

3. See Zhou SR 1986a, figs. 88, 89.

4. See Umehara 1943, 58; Izumi-shi 1984, pl. 64.

5. See Hunan 1981, 35ff, especially p. 35.

6. See Higuchi 1979, vol. 1, 244, for a discussion of inscription patterns and mention of two masters, Zhang and Chen. Under Zhang, the typical phrasing is as follows: "I, Master Zhang, made this mirror with exquisite skill. [I] shaped the likeness of immortals Wang Qiao and Master Chisong."

7. See Hunan 1960, pl. 78. This was made by Master Qinggai and found in an Eastern Han site. For another, excavated in Shaoxing, see Wang SL 1987, pl. 91. The Carter mirror presupposes this stage, as the tendency was to use cross-hatching in the hindquarters and to limit the rounded bosses to the lower edge only. The result resembles an octopus with bulging eyes.

43.

盤龍鏡

Mirror with a Coiling Dragon

Three Kingdoms (220–280)
Weight: 222 g
Diameter: 9.9 cm
Rim height: 0.8 cm
Overall height (knob): 1 cm
1995.323

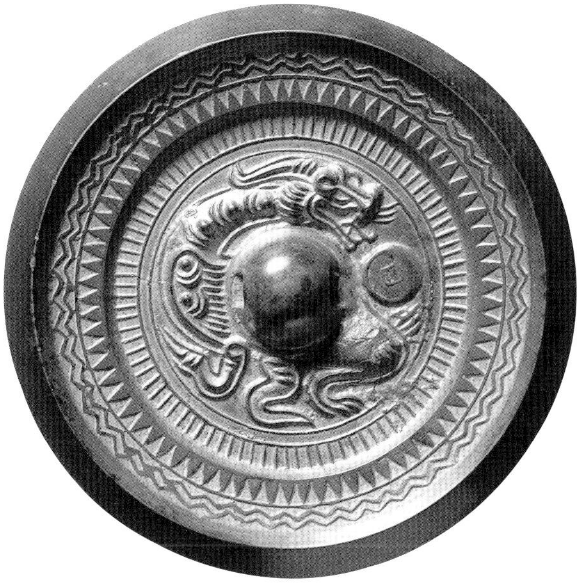

This is a smaller and simpler mirror than no. 42, but retains an enlarged knob, the deep well, a thickening rim, and a pronounced slant at its very edge. In the circle's inner well is a single dragon that coils around the knob and aims to catch the pearl that has materialized as a coin. The same sort of rounded bosses in mirror no. 42 also decorate its fore and hindquarters, though they are less articulated and assured. The stylistic concept, with a trifle more dissolution of the dragon form, may still be dated to the late Three Kingdoms, as there is insufficient reason for it to cross over to the Jin period (265–420). This type, it should be mentioned, began as early as Eastern Han.[1]

1. See Kong 1992, 477. The mirror was excavated in Guangzhou.

44.

四乳八獸鏡

Mirror with Four Nipples and Eight Animals

Western Jin (265–316), dated AD 300
Weight: 1,495 g
Diameter: 19.6 cm
Rim height: 1 cm
Overall height (knob): 1.3 cm
1995.334
(Plate 4)

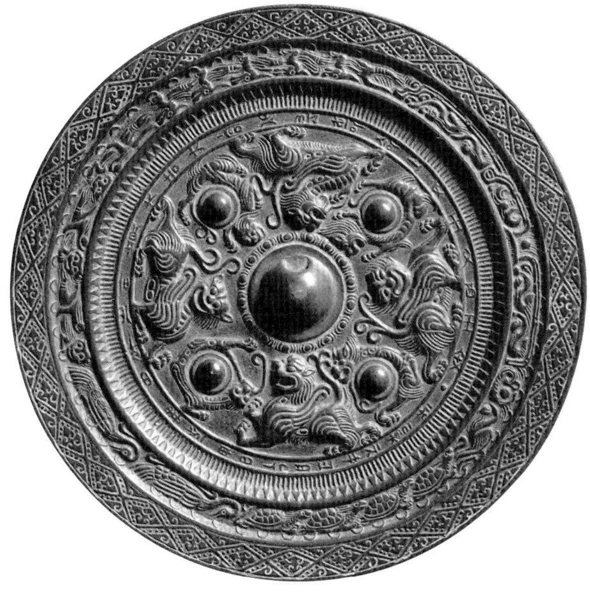

Dated mirrors are always important, as they set a chronological framework for dating other mirrors. This finely cast work is a rare instance of such a mirror in the Carter collection. The inscription states:

永康元年正月午日，
由(幽)[湅]黃日(白)。
早作明鏡， 大吉大宮(富)。
延長壽命，上有王父西王母。

On the day of wu, *the first month, the first year of Yongkang (AD 300),*
I made the bright mirror at the first of dawn
By refining the yellow and the white [metals].
Great fortune and wealth will come as a result.

May you live a long life.
Up above are the King Father [of the East] and the Queen Mother of the West.

As far as the Yongkang era is concerned, we favor not the reign of the Emperor Huan of Han (167–68), but rather that of Emperor Hui of the Western Jin, one which lasted only one year (i.e., 300). The latter is congruent with the style as exemplified in this and other mirrors made around that time.

One of the intriguing aspects is that, contrary to the inscription, the design on the mirror back contains neither the King Father of the East nor the Queen Mother of the West. One may attribute this to a careless attitude on the part of the metalsmith, who obviously was not a learned individual, and whose inscription was written in a manner that displays little calligraphic finesse. The decoration, by and large, is zoomorphic in nature. Around the hemispherical knob, four mythic *pixie* are counterpoised with four animals which, in an eccentric display, are pinned beneath the four nipples. One of the animals is obviously a monkey. The rest include a menacing bear, a fierce lion, and a hairy mammal. The design is intricate, and the metalsmith plays with contrasting raised threads, terraced effects, and double bands for the border. The outer band features interlocked lozenges accented with beads. The inner band is far more complex, displaying a recognizable theme (see no. 38). Counterclockwise, the thematic elements include six dragons pulling the chariot of the immortals, which in turn is followed by an unknown object to which an immortal is seen in obeisance. Behind him are four tortoises and two long-necked birds. Three tigers greet us at this junction, and then we are back again to the six dragons.

Despite this linkage with mirror no. 38, it may be worthy to note that this mirror no longer requires the rings of squares and semicircles. Its maker favored an expanded space where he could arrange the animals in extraordinary poses that defy tradition.

45.

瑞獸鏡
Mirror with Four *Pixie*
Northern and Southern Dynasties (420–589)
Weight: 347 g
Diameter: 11.6 cm
Rim height: 0.6 cm
Overall height (knob): 1.3 cm
1995.327

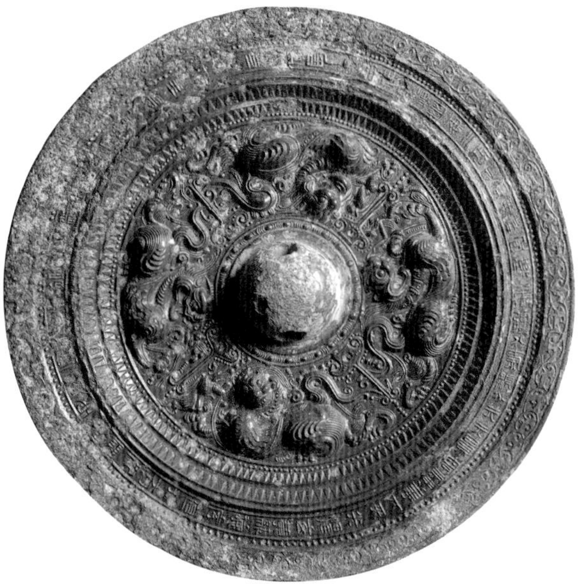

Corrosion has damaged a portion of this mirror's surface including the inscription, which can be read only partially:

吾作明鏡，幽湅三商。
雕刻典祚，配德萬疆。
人氣顯分，天元景容。座事六采，
光耀玄黃。統德序道，□有紀綱...

I made this bright mirror
Having melted three metals.
I have sculpted the classic paradigms
And the virtuous in myriad years.
The human spirit is made manifest
And heavenly deities reveal their noble forms.
As they are painted with six colors,
Their light shines through the heaven and earth.
With virtues and the Way in grasp,
Rules and laws sprang into being . . .

The decoration on the mirror back displays two pairs of confronting *pixie*, with a central candelabra in between individual animals. Smaller figures are also seen in the head portions of these mythic animals, but not in the tail quarters. These winged *pixie* are powerfully shaped. Despite the images' small size, their torsos and hindquarters bulge with the same sort of grand rhythm seen in, for instance, the *pixie* guarding the tomb of Xiao Xiu, Prince Ancheng of the Liang dynasty (d. 518).[1]

Indeed, in its careful articulation of the bodily form, in the clarity of the quadrant structure, and in the stateliness of presentation, this mirror has clearly rejected the earlier predilection for the unusual and the eccentric, as exemplified by mirror no. 44. It is a virtual prelude to the coming era of the Sui and Tang, when these qualities became paradigmatic.

1. See Chu 1935, figs. 41, 42. When compared to mirror no. 40, these *pixie* begin to lift themselves from the background with conviction.

Phase III

Sui and Tang Dynasties 581–907

46.

纏枝花十二生肖鏡
Mirror with Interlacing Floral Motifs and Twelve Zodiac Animals
Sui (581–618)
Weight: 294 g
Diameter: 15.4 cm
Rim height: 0.45 cm
Overall height (knob): 1.4 cm
1995.336

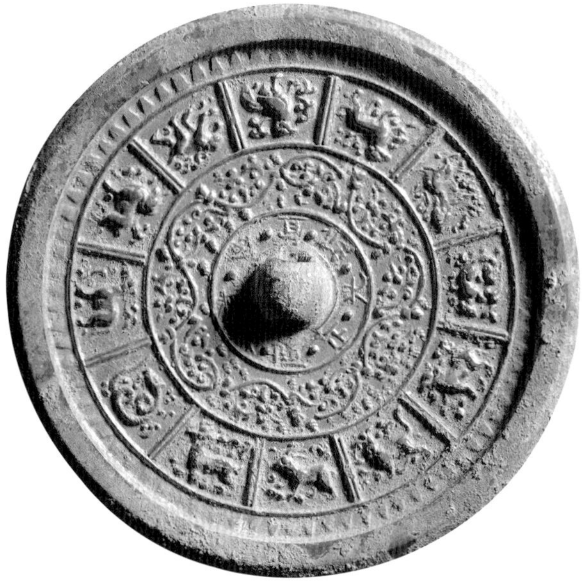

Clearly the images on this thin bronze mirror were built from wax, but the casting lacks precision. The design consists of a central knob and a series of concentric circular bands. The innermost band contains an eight-word inscription, with phrasing typical of earlier mirrors found in the Sui and early Tang periods:

光正隨人，長命宜新。

This mirror's light shows the man inside.
May you live a long life and [constantly]
renew yourself.

Floral branches interlace in the second band. The controlled undulation produces five segments. The third band is divided into twelve parts, which house the zodiac animals in the clockwise sequence: rat, ox, tiger, hare, dragon, snake, horse, sheep, monkey, cock, dog, and boar. A sunken ring of the well-known sawtooth design continues to be effective, before the undecorated rim sets the final enclosure.

Sui provenance for this type of mirror receives ample support from archaeological finds, including a mirror from Li Jingxun's tomb in Xi'an, dated to the fourth year of Daye (608) during the reign of the Sui emperor Yangdi.[1] This find might also suggest a pre-seventh century date for the type, as buried mirrors were often cast decades prior.

1. See IOA 1980, pl. XII-(1). Additional examples can be seen in Shaanxi 1959, figs. 87, 88.

47.

重圈神獸鏡
Mirror with Deities and Animals in Concentric Circles
Sui (581–618)
Weight: 1,393 g
Diameter: 33.4 cm
Rim height: 0.9 cm
Overall height (knob): 1.6 cm
1995.337

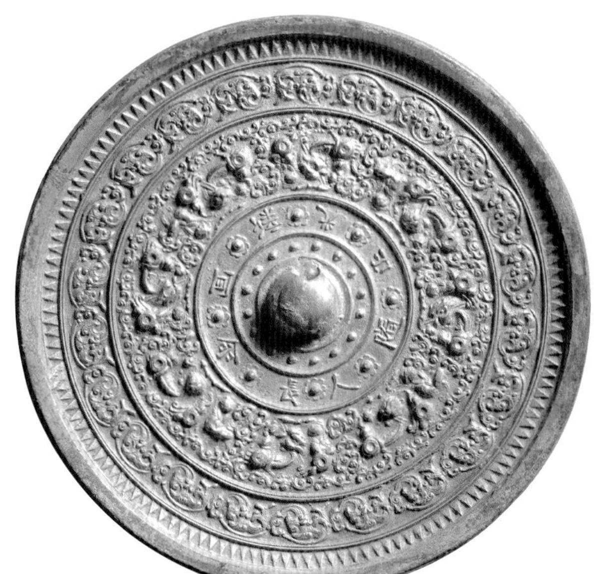

This large mirror with silvery luster shares with *Mirror with Interlacing Floral Motifs and Twelve Zodiac Animals* (no. 46) the same message in its inscription:

A mirror's light shows the man inside.
May you live a long life and [constantly]
renew yourself.

The calligraphy, however, is far more accomplished. It is a style that retains a touch of the clerical script, but is also on the verge of approaching the *kai* (regular) script.

These aspects partially affirm a Sui dating for this piece. The concentric-circle design emphasis reveals a total of five bands, which is rare. From excavated examples, the closest parallel found is a mirror from a Sui tomb to the west of Xi'an, but in that case the concentric bands number only four.[1]

Despite the numerical difference in bands, these two mirrors share a number of other features, including the circle of beads in the vicinity of the knob, and in the outer bands the interlocking fish as well as the sawtooth motifs. Both mirrors contain deities and animals in their widest inner band, but differ considerably in the manner of presentation. The Xi'an mirror allows the lithesome forms of the animals to sweep across the vista, giving the decoration a swift pace and momentum. The Carter version favors a gradual rotation with a series of smaller images, circular rings, and trailing clouds, which are replicated no less than sixteen times.

These last elements also link the mirror to an Eastern Han antecedent (see no. 38). The latter is smaller in size, but the figures and *pixie* are larger, proportionately speaking, and well articulated. In the present mirror, the increased complexity is adversely combined with the lack of precision in casting. The result is blurred images and the loss of individual identities. Because of this we can discern only a number of *pixie* with scrolls in their jaws, seated deities either over the circular rings or suspended in space. If the divine pair of King Father and Queen Mother is still present, we are not able to pinpoint them precisely.

Thus, the Carter mirror is perhaps not as daring and innovative as the Xi'an example. Its tie with the past is apparent, even though its maker made a valiant attempt to explore the concentric-circle design and to enlarge the sphere of reference, in content as well as in size. Because of this, it may still be considered an ambitious undertaking.

1. See Shaanxi 1959, pl. 85

48.

方形掌中鏡
Miniature Square Mirror
Sui (581–618)
Weight: 42 g
Dimensions: 3.8 x 3.8 cm
Rim height: 0.5 cm
Overall height (knob): 0.6 cm
1999.219

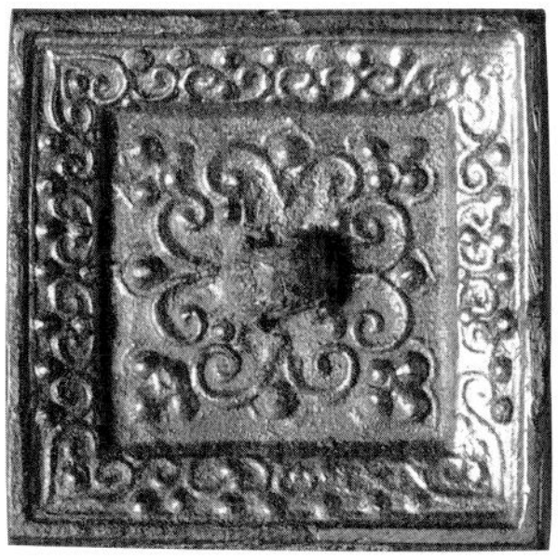

In the design of a miniature mirror, achieving both charm and stateliness is no mean feat. This is exactly what the maker of this mirror has done. Within the simple square format, curves and round drops replicate themselves either in a central eight-petal blossom or, along the sides, in flower buds. Along the sharp cut of the frame-like border, the curves are repeated in a narrow row, with round drops that align at the very edge. A heart shape occupies each of the four corners.

The result is a design of remarkable complexity and clarity that also yields endearing cycles of tension and relaxation, expansion and contraction. Without visible effort, this mirror draws us into its own heart and soul.

49.

四神鏡
Mirror with Four Spirits
Sui (581–618)
Weight: 870 g
Diameter: 19.8 cm
Rim height: 0.65 cm
Overall height (knob): 1.5 cm
1995.339

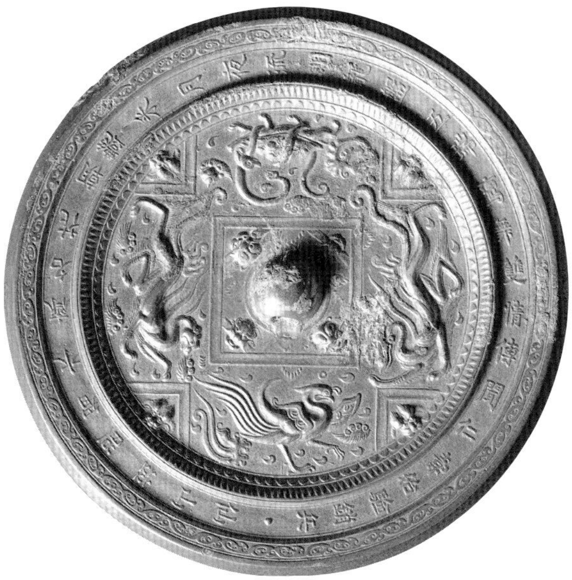

This mirror type is an apparent reworking of the old TLV pattern. Here, the T's and L's have vanished, but the combination continues of the central square and the V's which demarcate the four spheres, which in turn house the Four Spirits, the familiar set of Blue Dragon, Red Bird, White Tiger, and Dark Warrior (of snake and tortoise intertwined). Whereas the old design gives rise to a newer version, it consists largely in a tectonic structure and a graceful portrayal of these animals in isolation, accompanied only by the trailing clouds. In addition, the rhythmic pulse here appears more as an extension of the spirits' lithesome bodies, rather than as the result, as in the past, of the overall pattern. The appearance of the band of inscription, written in the true *kaishu* (regular script) in a poem of four words in a line, is equally notable, as it reflects the state of calligraphy as well as the literary penchant of the time. The mirrorsmiths of the Sui and Tang periods, and presumably their clients, did not tolerate casual abbreviations and incomplete messages as was seen in Han period mirrors. The inscription reads:

仙山並照，智水齊名。
花朝艷采，月夜流明。
龍盤五瑞，鸞舞雙情。
傳聞仁壽，始驗銷兵。

The immortal peaks rise as twins.
The wise river now has a rival.
Morning flowers are made bright and charming.
At night, its light reflects the moon.
The dragon coils around the jade disc.[1]
A pair of fabulous birds dances in harmony.
It is said the Renshou era (601–4)
Was when the war began to end.

For dating purposes, besides the two examples published by Umehara in *Tokyo Taikan*[2] dated to the first year of the Tang dynasty's Yunhui era (650), we need to consult archaeological finds in China. A similar mirror with four animals carries a *terminus ante quan* of 611 and is found in a tomb site in Xi'an.[3] Sui and early Tang tombs at Changsha, Hunan province, also have yielded close parallels.[4] This distribution extends to the north, as a Tangxian site in Hebei also produces a piece with an identical inscription (or part of that inscription).[5] Thus, even if this mirror type could have survived into the early Tang period, where the Four Spirits began to acquire weight and volume, it is clearly of Sui inspiration. The slender proportions of these Spirits in the Carter mirror also reveal its Sui heritage.

1. The term *wurui* refers to five different jade discs which, by imperial decree, are reserved for five official ranks. Rather than translating it literally, it would be best to retain the idea of a single dragon coiled within the circular forms of the mirror.

2. Umehara 1984b, pls. 8, 9. The Yunhui mirrors display the Four Spirits in a way that shows them to be heavier in weight and proportion, versus the lithesome configuration found in the Carter mirror.

3. See Shaanxi 1959, pl. 81, also pls. 82–84.

4. See Hunan 1960, pls. 89–91; Kong 1992, 499–501.

5. See Hebei 1996, pl. 92. This mirror is incomplete, and part of its inscription has disappeared. Its diameter measures only 16.7 cm, smaller than the Carter version.

50.

瑞獸鏡
Mirror with Four Running Animals
Tang (618–907), early 7th century
Weight: 156 g
Diameter: 9.6 cm
Rim height: 0.7 cm
Overall height (knob): 0.9 cm
1995.338

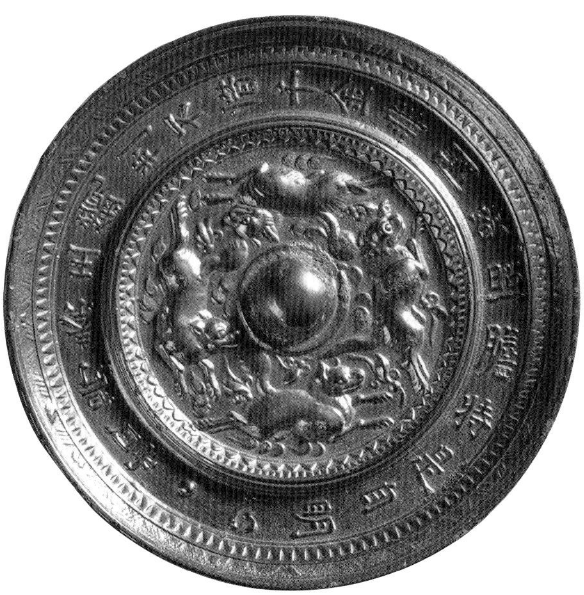

When compared with *Mirror with Four Spirits* (no. 49), the elimination of the "square" surrounding the knob helps to create an unmitigated, concentric circular platform. The animals here, whose identities are at best ambiguous and somewhere between a dog and a wolf, run in a counterclockwise fashion in the inner band. The outer band prominently features the inscription, a poem in the quatrain mode:

賞得秦王鏡，判不惜千金。
非關欲照膽，特是自明心。

I received a gift: the mirror of the King of Qin.
For this, I would have paid willingly a thousand taels of silver.

In no way would I use it to unveil your hidden desires.
Verily, I only wish to lay bare my own heart.

Legend has it that in the Xianyang Palace of the King of Qin, a large mirror was found that had the capacity, like an x-ray machine, to reveal the inner organs of those who stood before it.[1] The same legend also says that if a woman with licentious thoughts should come before it, her gall and heart would be shown in an inordinate state of palpitation. Such was the effectiveness of this mirror that the first emperor of China, Qin Shihuang, had a number of his palace ladies executed. While apparently alluding to this legend, the author of the poem thought to subvert its cruel intention and instead skillfully create a distinction between the subject of love and the lover. The owner of the mirror, the mirrormaker cleverly intimated, wished only to reveal his own soul, and did not doubt the hidden desires of the lady fair.

The typology itself—specifically the running animals in the inner band around the knob and surrounded in the outer region by the *kaishu*-style inscription—is clearly of Sui/Tang heritage. The animals' volumetric form supports a date of early Tang.

Preserved in excellent condition, with the silvery luster of *shuiyinqin*, it is a small mirror with a thickened plate. Despite its size, the significance of this type of mirror is clear. This is the base upon which the "animal-and-grape" typology was founded.[2]

1. I took the liberty of translating the word *dan* (gall) as "hidden desires." For this legend, see Ge 1929, *juan* 3, 3a, 3b. Some scholars prefer to associate the King of Qin with Li Shimin, i.e., the Tang emperor Taizong, who prior to his assumption of the throne was awarded that princely title. This is not warranted at all.

2. See Xu D 1994, 263–342, especially p. 304.

51.

蓮座瑞獸鏡
Mirror with Three Pairs of *Pixie* around a Loti-form Knob
Tang (618–907), mid 7th century
Weight: 989 g
Diameter: 17.6 cm
Rim height: 1.3 cm
Overall height (knob): 1.5 cm
1995.335

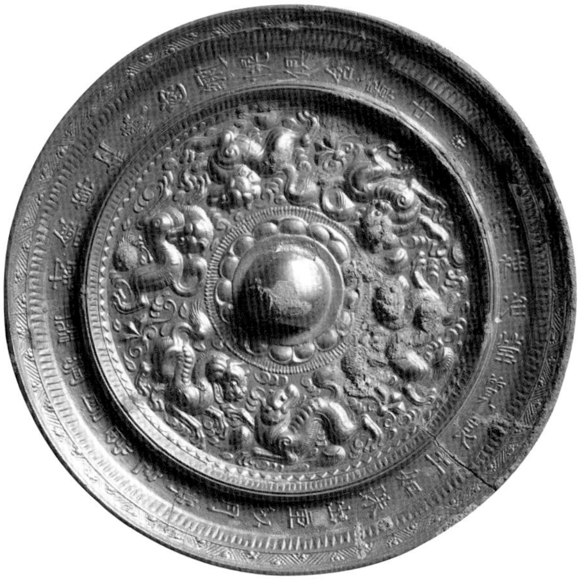

A variant of *Mirror with Four Running Animals* (no. 50) and larger in size, this mirror is not as well preserved. Patches of corrosion mar the speculum surface, and a notable crack is present. Barring these flaws, the casting is precise, with the central knob surrounded by sixteen petals so as to assume the appearance of a lotus blossom. In the inner circle, the three pairs of *pixie* are displayed in fuller relief and sense of organic coordination than animals in Sui period mirrors (see, for example, no. 49), which are slender in proportion and lithesome in appearance. This may be an indication that this mirror type represents the Tang dynasty style proper. A nearly identical mirror was found in the Xi'an region,[1] but with a different inscription from that here:

鍊形神冶，瑩質良工。如珠出匣，似月停空。當眉寫翠，對臉傳紅。綺窗繡幌，俱含影中。

The spirit merges with the forms that are being wrought.
The lustrous metal matches the exquisite craftsmanship.
Like a pearl, it emerges from the container box—
A virtual moon suspended in the firmament.
[It enables] the lady to draw her eyebrows
And apply rouge to her face.
Even those elegant windows and the embroidered curtains:
Can all be reflected within.

Scholars have linked this poetic inscription to Wang Yan 王衍 (r. 919–25), the ruler of the former Shu.[2] If so, mirrors bearing this inscription, for which there are several examples, must postdate Wang Yan's short life (he died on 28 *sui*). However, mirrors with this very same inscription were being produced during the Sui as well as the early Tang periods, which demonstrates that this linkage was ill conceived.[3]

1. See Shaanxi 1959, pl. 103.

2. See Dong et al. 1962, vol. 3, 1629.

3. See Kong 1992, 515, which publishes one such example from a tomb of Sui or early Tang pedigree.

52.

瑞獸葡萄鏡
Animal-and-Grape Mirror
Tang (618–907), early to mid 7th century
Weight: 576.5 g
Diameter: 11.7 cm
Rim height: 1.2 cm
Overall height (knob): 1.3 cm
1995.340

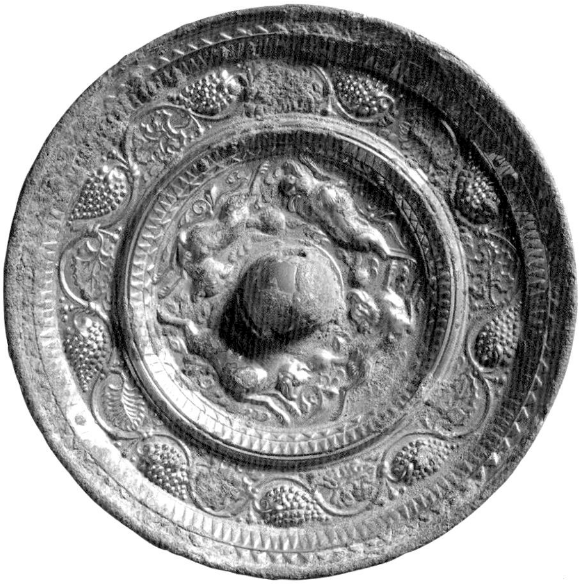

While the state of preservation for this particular mirror is less than ideal, it nevertheless provides an idea of how the animal-and-grape design, a major typology, came into being. The interior portion is essentially identical to mirror no. 50, where four animals run in a counterclockwise fashion around the hemispherical knob. In the outer band the inscription has vanished, a victim of time. In place of the inscription is a motif of undulating grapevines that comes to assume a greater importance. The vines arch over the grape clusters and dip under the leaves before rising again, creating a regulated sequence of alternating images in a wavy pattern.

A dating to the first half of the seventh century is plausible for this piece, when compared with a mirror-fragment unearthed from Xi'an, tomb no. 568, dated 665.[1] The casting of this mirror must have preceded the burial by decades, placing its creation in the first half of the seventh century. Notable too is the introduction of grape bundles in this Xi'an mirror, in contrast to their complete absence in the interior of the Carter mirror. The Xi'an mirror also incorporates such motifs as long-tailed birds in the border, which heralds a further expansion of the decorative repertory. Those decorative elements are not evident in the Carter version. Thus, it seems plausible that the Carter mirror in effect was standing at the very inception of this new typology. We may set the second quarter of the seventh century as the time of its creation.

Grapes were introduced into China as early as the Han dynasty and soon appeared in textiles as a decorative motif. During the Tang period, interest in this imported fruit deepened considerably and grape wine gained enormous popularity in the court and among the gentry.[2] This interest made its way into mirror design, with the typology of animal-and-grape as its most eloquent testimonial. The mirror type itself did not last long, yet at its peak, its vitality was irrepressible and its wealth of details unprecedented.[3] At its inception, however, it was notably simple and subdued, as attested to in this Carter mirror.

1. See IOA 1966, pl. XLI, no. 3.

2. See Yen 1989, 302–3, for a brief introduction to the history of the grape and its importation into China.

3. Ibid., 312–15. Yen believes that the time span for this type lasted from 650 to 710.

53.

瑞獸葡萄鏡
Animal-and-Grape Mirror
Tang (618–907), mid 7th century
Weight: 215 g
Diameter: 8.2 cm
Rim height: 1 cm
Overall height (rim): 1 cm
1995.356
(Plate 5)

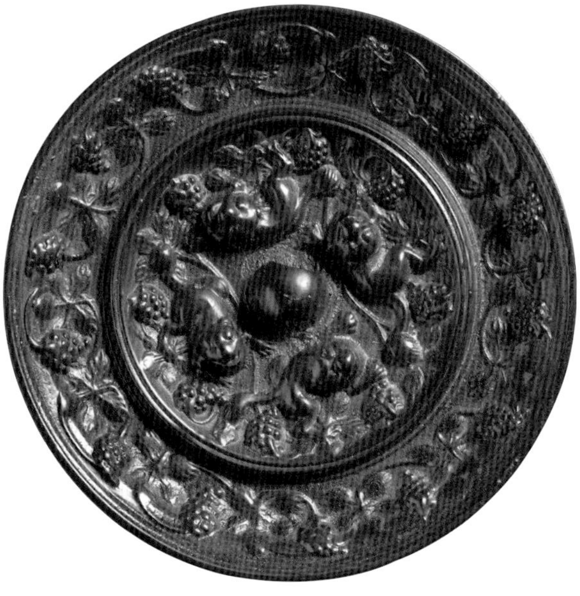

This small mirror with the dark, luminous patination known as *heiqigu* is in excellent condition. Within the thick rim a clear division between the interior portion and the border can be seen. In the inner circle surrounding the knob, the animals assume a variety of crouching poses over a field of vines and grapes, while looking upward as though at a viewer.

In the outer band, the grape plants take control, with alternating leaves and fruits in a dense formation. Perhaps due to the small size, the mirrormaker may have sought to limit the repertory of motifs, and there are no avian presences to complicate the matter.

Based on the presence of the grape bundles in the inner zone, and on the absence of any long-tailed birds, we date the mirror to the mid seventh century.

54.

葡萄紋掌中鏡
Miniature Mirror with Grape Decoration
Tang (618–907), 7th century
Weight: 61 g
Diameter: 4.9 cm
Rim height: 0.7 cm
Overall height (rim): 0.7 cm
1999.218

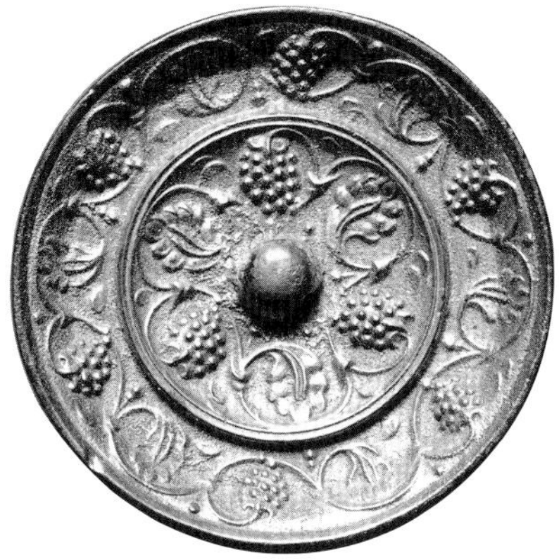

As small as this mirror is—it can certainly be held in one's palm—the execution is simply superb. The concentric arrangement is faultless as it revolves around a small hemispherical knob. In the interior, animals have disappeared, with only the grapevines to string along the leaves and fruits and form what is roughly a hexafoil, with three clusters of grapes and three leaves in alternation. In the outer compartment, the same pattern of alternating leaves and grape clusters also prevails. Careful scrutiny reveals that six modular units form a chain. Each unit begins with a vine that grows out from a previous one and then sprouts two branches. One of the branches arches backward and bears a drooping leaf, while the other makes a full swing until it comes to a junction where it in turn sprouts two additional branches. One produces a grape cluster and does so by going under it and the other then moves forward to begin the whole process again. The direction may be counterclockwise, but it also has a contrary pull. The rhythm is intricate and varied, entailing a circular revolution and also a push and pull, an expansion and contraction. Even the images themselves are an integration of opposites: softly curled leaves versus solid bundles of grapes. The mirror may be small, but it is the "mustard seed" that enables us to see the world of art within.

55.

瑞獸葡萄鏡
Animal-and-Grape Mirror
Tang (618–907), late 7th century
Weight: 1,097 g
Diameter: 14.8 cm
Rim height: 1.45 cm
Overall height (rim): 1.45 cm
1995.353

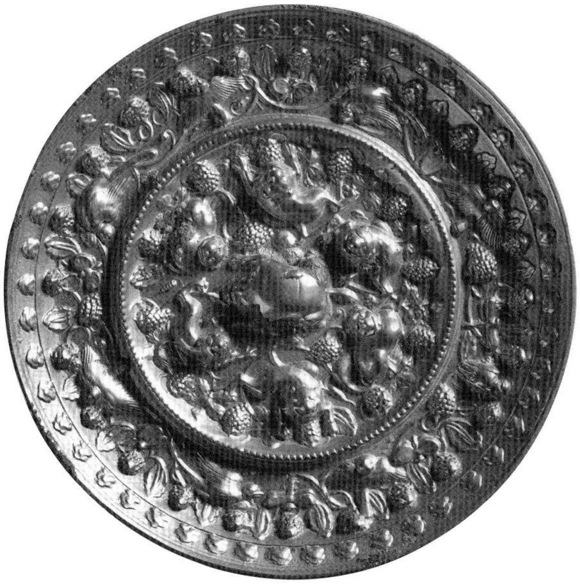

Minor corrosion mars the edge of this mirror, which is otherwise in a fine state of preservation. Comparison with mirror nos. 52 and 53 shows that this mirror not only has gained in size and thickness (and the rim displays a greater height than that of the knob), but the design has become increasing complex. The knob now assumes the form of a crouching beast. Around it are six more animals assuming three sets of poses, each aligned axially at opposite ends. Two are seated in profile with heads turning directly toward the viewer. Two others angle their torsos and heads toward the zoomorphic knob. The remaining two turn away from it by facing outward. The grape plants remain within as well as without the inner circle. In the border, a variety of motifs are evident. Among the birds, with long or short tails, there are those rendered in profile and in flight; they are seen from above, with extended wings, and at least one assumes a standing pose with feet pointing downward to the mirror's center, thereby reversing the viewing order of its brethren. Insects are represented as the bee, dragonfly, and butterfly. Close to the rim, forty-eight small blossoms circulate around the mirror. The overall pattern is one of clear-cut division between the interior and the border, with a burst of highly articulated images that spread across the mirror exacting both centrifugal and centripetal forces. The alloy's luster also helps to reflect light where the relief is high, with an impressive play of sharp contrasts and sudden shadows.

A comparable example is the famed mirror from the tomb of Dugu Shizhen, dated 698, near Xi'an. The Dugu version is larger (diameter: 16.85 cm).[1] Yet a basic affinity remains in the clarity of division between the outer and inner spheres, as well as in the lushness of the imageries. Again, taking into consideration the difference between the time of casting and the object's burial, the Dugu Shizhen's mirror could very likely be dated in the third quarter of the seventh century. We shall do the same with the Carter version.

1. See Wang ZS 1982, 410. This is a revised measurement; an earlier report in IOA 1980, 29, reported the diameter to be 16.9 cm.

56.

瑞獸珍禽葡萄鏡

Animal-and-Grape Mirror with a Variety of Birds

Tang (618–907), late 7th century
Weight: 741 g
Diameter: 13.9 cm
Rim height: 1.4 cm
Overall height (rim): 1.4 cm
1995.355

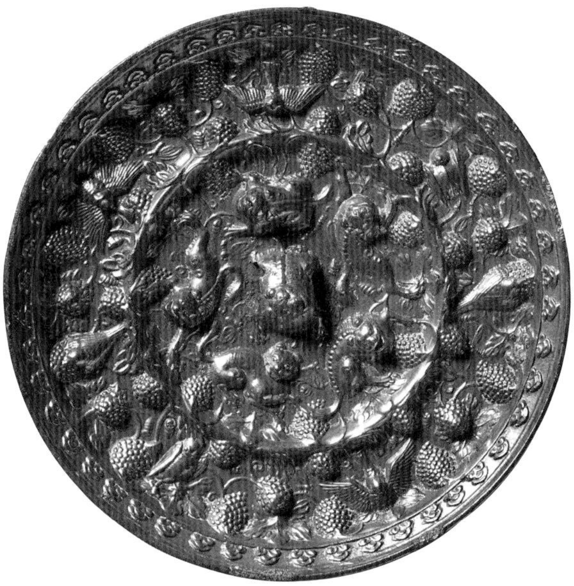

As a fine example of the animal-and-grape typology in the Carter collection, this mirror is similar to no. 55. Its knob assumes a crouching zoomorphic form. The animals in the inner circle are in varied postures, and the grapevines have invaded this circle as well as becoming the main motif in the outer sphere. This allows the birds to appear amid the vines' lush foliage and crisscrossing tendrils. Given the tangled vegetation, the birds can assume only a limited range of movement. They either spread their wings, stand still by attaching themselves to the vines, are portrayed from top down, or, in just one case, pose sideways. Forty-nine cloud-floral motifs are evenly distributed at the rim's edge.

The animals have now acquired a series of small, bead-like, bony protuberances at their spines—accentuating their posture and creating echoes from the grape bundle motifs, which are similarly beaded. Spreading the bead pattern across the divergent motifs enhances the mirror's cohesiveness.

Partly to reinforce the fusion of this design, the grapevines begin to climb over the circular ridge that once separated the inner from the outer compartment. This act of crossing over the boundary is important: in the context of the animal-and-grape typology, it signifies a further stage of development beyond what was reached in mirror no. 55. A similar indication also comes from another mirror, one that was found in the combined burial of General Anpu 安菩 (d. 664) and his wife (d. 704), dated in the year 709.[1] The mirror was part of the wife's tomb furnishings,[2] and its production must have preceded the burial by a decade or two. We are therefore compelled to date Madame Anpu's mirror—and the Carter version as well—to the last decades of the seventh century.

1. See Luoyang 1982, 21–26, pl. 7-(1).

2. See ibid., 22, fig. 2, for the plan of the tomb and the respective funerary furnishings. The report is noncommittal with respect to where each member of the couple was buried, as the burial site contained two chambers. A comparison between the grouping of pottery figures should be instructive. In the eastern chamber are nine male figures and two female ones. In the western chamber, the ratio is reversed: six males and eight females. The latter is likely to be the wife's quarters in the netherworld; it is also where the mirror was found.

57.

寶相花鏡
Mirror with Six Circular Flowers
Tang (618–907), 7th century
Weight: 833 g
Diameter: 16.6 cm
Rim height: 0.9 cm
Overall height (knob): 1.5 cm
1995.341

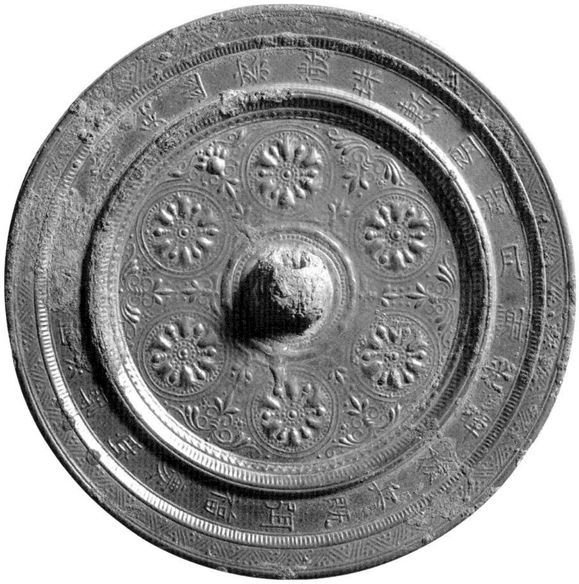

This mirror and its brethren can be found in the centers of Xi'an and Luoyang, as well as in Yangzhou and elsewhere.[1] Characteristic is the classic pattern of the inner well and outer border, made definitive by a full circular ridge in high relief. The inclusion of a ribbon of inscription in the outer sphere also testifies to an early date.

The design is rich and stately, carrying the ethos of the Sui period to the early Tang. Around the knob are six floral patterns encased within concentric circles. Helping to assign them into six sectors are candelabra-shaped dividers. The floral images here seem to rise subtly from the ground, and emerge in clarity only at the end of the process. They are in stark contrast against the mirror's solid and unyielding surface.

Long after the mirrors were made and their makers vanished into history, these floral motifs became known as *baoxiang hua*, "flowers of precious appearance," instantly evoking a Buddhist linkage. When analyzed with a degree of care, *baoxiang hua* is but a general term, applicable to those flowers that, beautiful as they are, clearly defy identification. In this volume, we will use the rubric without its religious connotation.

The inscription reads:

□□□ 膽，知幽察微。珠慚朗潤，
月謝光輝。□兵既弭，福慶斯歸。

[This mirror can unveil one's] gall,
It can fathom the hidden and the subtle.
Its clarity and luster put the pearl to shame,
And the moon to pale.
[Now] that the war has ended,
Let's celebrate the rising fortune at the wake.

A casual glance at existing compilations of inscriptional data shows this six-line verse is a rare occurrence. In our limited experience, we have not seen any other mirror bearing this inscription. This warrants a claim for its uniqueness.

1. See Kong 1992, 593. For Lüshun finds, see Lüshun 1997, pls. 89–90.

58.

寶相花鏡
Lobed Mirror with Six Blossoms
Tang (618–907), mid 7th century
Weight: 961 g
Diameter: 19.1 cm
Rim height: 0.55 cm
Overall height (knob): 1.2 cm
1995.360

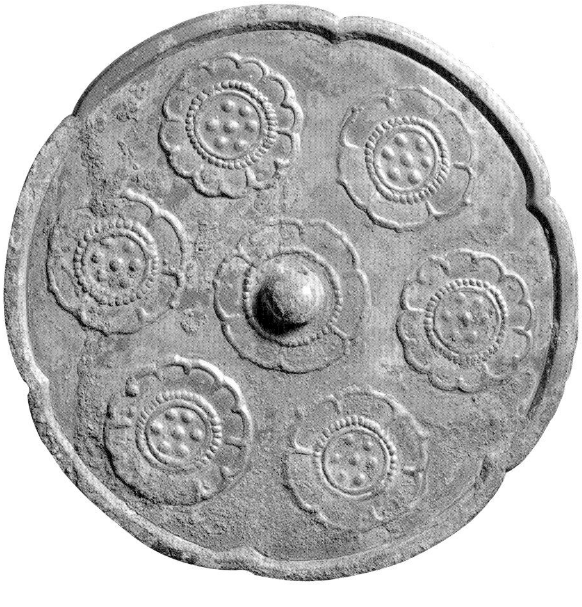

This mirror—the so-called *kuihua* variety—has a lobed contour in six arcs. The advantage of this type is that it allows a simple, undivided surface. The increase in space can provide its maker with sufficient freedom to explore the full extent of the design.

A small hemispherical knob on a six-petaled floral base becomes the center around which six additional blossoms are evenly distributed. Three of the six have, just as the central flower, six petals, while the rest have twelve petals. Judging from other surviving examples, those blossoms with six petals are fairly constant, whereas the others can take on different forms and change from mirror to mirror. In any case, a common feature for all of the blossoms on this mirror is the pod within, each displaying seven seeds. Just like the gestalt of the mirror itself, one seed is at the center, with six other seeds revolving around it. The exception is the central blossom, whose center is occupied by the knob.

For comparison and dating purposes, nothing can match the mirror from a Tang tomb in Xi'an, with the burial dated to 670.[1] From this mirror we learn that the *kuihua* type was already in existence. We are also able to surmise that the particular approach to mirror design described above had already been successfully tried. With such double persuasion, a mid seventh century date would seem both reasonable and compelling.

1. Shaanxi 1959, pl. 150. It is from a tomb dated to 670, with the mirror made perhaps earlier. It is also the earliest known, datable example of a *kuihua* mirror. Also see Zhou/Zhou 1979, 53–58, figs. 3, 11. In Chen P 1987, 58, explanatory note on pl. 97, the author also mentions a Liaoning find of the early eighth century.

59.

花卉葵花鏡
Lobed Mirror with Eight Blossoms
Tang (618–907), 8th century
Weight: 1,493 g
Diameter: 24.5 cm
Rim height: 0.7 cm
Overall height (knob): 0.9 cm
1995.359

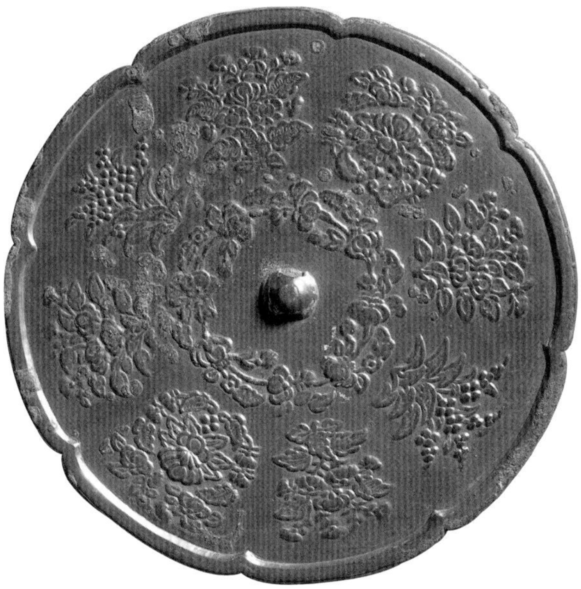

This is a large mirror in the *kuihua* shape with a small knob surrounded by a unified field, which is used to full advantage. Around the knob is a wreath of flowers, buds, and leaves. The main motifs lie away from the wreath, consisting of eight separately rendered blossoming plants. Though the plants may be eight in number, there are only four varieties. Each of the four is repeated across the axis, thus forming pairs of peony, camellia, begonia, and gardenia, all enclosed within the lobed contour of the mirror's rim. As far as floral mirrors are concerned, this and the mirrors like it stand at the very apex of the genre, during the proverbial High Tang of the eighth century.[1] While years of burial did some damage to the surface, it did not dim the mirror's inherent richness and opulence. In its full and unmitigated devotion to flowers, it enables us to perceive the passion of Tang men and women toward the world of nature in its most resplendent manifestation.

1. A mirror with six clusters of blossoms is found in a Zhengzhou burial of 797. See Zhengzhou 1995, 23–39, especially p. 27, fig. 11.

60.

雀繞花枝鏡
Mirror with Sparrows and Floral Sprays

Tang (618–907), late 7th–early 8th century
Weight: 191 g
Diameter: 9.3 cm
Rim height: 0.8 cm
Overall height (knob): 0.8 cm
1995.358

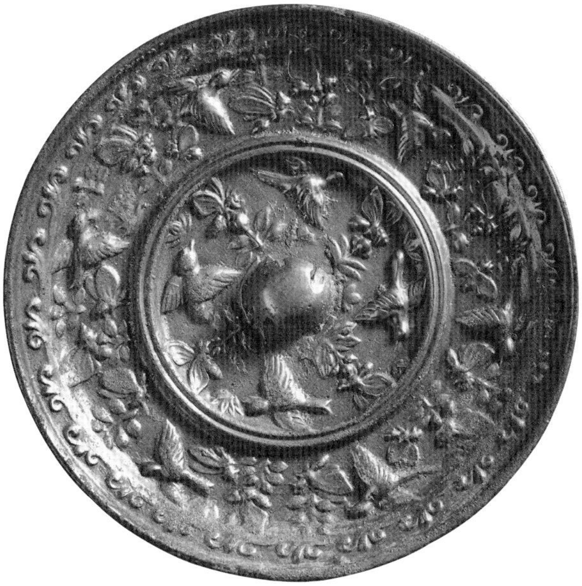

While its casting is less than salutary, the mirror here is close to, or nearly identical with, two archaeological specimens: one found in a tomb at Yanshi, dated 706,[1] and another from a tomb near Xi'an, dated 714.[2]

Given that these are the dates for the burials, and not for the mirrors, we may suspect a late seventh century chronology as appropriate with respect to this type in general, and the Carter specimen in particular. The mirror has a classic arrangement that includes a round knob, and a strong ridge of circle that divides the mirror into the inner well and the outer band. This typology is akin to mirror no. 50, where the four animals make a run around the knob. Here one finds four sparrows in flight, with bees and butterflies that alight on the tips of trailing plants that serve as dividers. In the outer realm, the same motifs—birds, insects, and plants—are repeated, expanded, and varied. At the very edge, a sequence of tendrils (in triplets) encircles the mirror.

Together with the previous floral mirrors (nos. 57–59), we sense the formation of a trend that would take away some of the importance from zoomorphic images and which would allow for an expansion in other genres and interests. With this mirror, the avian presence has now increased in range and intensity. We may place this type of mirror under the general heading of "flower-and-bird."

1. See Xu D 1994, 312, fig. 5-(1), pl. 2-(3).
2. See Shaanxi 1959, pl. 140. For another example, see IOA 1966, pl. XLII-3; the site, however, is undated.

61.

金銀平脱花鳥鏡
Mirror with Flowers and Birds in Gold and Silver Cutouts over a Lacquer Base

Tang (618–907), mid 8th century
Weight: 377 g
Diameter: 16.8 cm
Rim height: 0.35 cm
Overall height (knob): 0.9 cm
1999.217

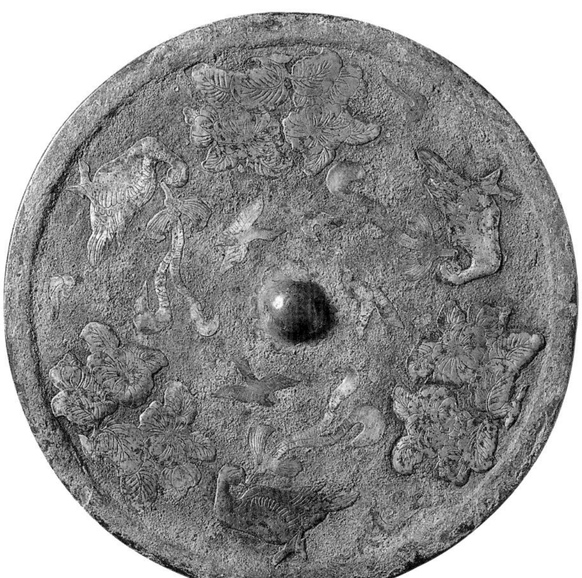

Four sparrows or magpies circulate around this mirror's small knob. Beyond them, wild geese alternate with paired plants. From their beaks, the geese display a twin tassel tied in a knot at the base which then surges upward into separate strands, a motif seen elsewhere.

Technically, this mirror is decorated by the use of silver and gold cutouts—with further engraving to clarify the floral and avian images—laid over a lacquer base. As the receptacle of the lacquer juice, the mirror's back has a raised rim to contain the outflow of this natural "plastic." However, given its antiquity and the adverse burial conditions, lacquer layers have disappeared, leaving the cutouts in only a fair state of preservation. Typical of this mirror type, the knob is small to allow optimum surface area for the layout, which can be as simple as the example here, with its larger motifs that cover the surface, or as delicate and complex as the finest example found in the CMA's collection (1973.74).[1]

This highly involved and complex process was not limited to mirrors, but was extended to a variety of luxury products. Court production during the time of Emperor Ming was notable, and the emperor was known to have bestowed such objects as gifts to meritorious officials, including the infamous An Lushan, who led the rebel army against the imperial reign. In later times, the Tang court sought to halt this practice, once in 757 and again in 772.[2] Under such pressure, the production of *pingtuo* mirrors, like other objects made by the same technique, began to decline.

1. See Kong/Liu 1994, 128, where the authors create an artificial category of 貼金貼銀鏡 (pasted gold and silver cutouts) and attempt to separate it from the usual *jinyin pingtuo*.

2. Ibid., 124–27.

62.

神仙山岳菱花鏡
Octafoil Mirror with Riding Immortals and Rising Peaks
Tang (618–907), 7th century
Weight: 334 g
Diameter: 12.2 cm
Rim height: 0.75 cm
Overall height (rim and knob): 0.75 cm
1995.344

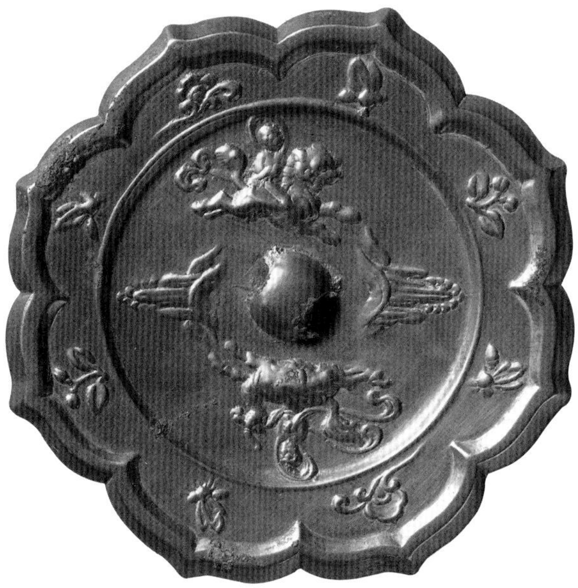

Characteristic of the Tang *linghua* mirror, this example has eight arched petals and a circle within. In the inner compartment, the knob is the core from which riding deities and mountains alternate in a revolving movement. The eight border compartments contain four different sets of motifs: a butterfly, a spray of flowering plant, a bee, and a floating cloud. The two members of each pair face one another across the axis.

For dating purposes, we may need to consult excavated or datable examples where immortals riding on animals or birds appeared. An identical example was excavated in Xi'an, Shaanxi province.[1] However, apart from being a Tang work, this mirror fails to help us set up a narrower temporal framework. More useful is the bronze mirror found in Yan Wan's tomb as reported in *Wenwu*.[2] Yan Wan, the eldest daughter of Yan Lide (d. 656), married into the imperial family and subsequently died in 690. She was buried by the side of her son, Li Hui, in 724 in Yunxian, Hubei.

The royal connection, as well as the heritage of Yan Lide, the most notable craftsman/artist in the imperial service during the early Tang period, furnished the context for the mirror in the tomb. This mirror is in a *linghua* shape with a series of border motifs identical to those in the Carter mirror. Unlike the latter, it has four immortals in the inner circle, riding on animals and phoenixes—but without mountains. That this was a popular type can be attested to in Shōsōin as well, which possessed three examples in its Southern Storage.[3] The figures in the mirrors are very similar among the Yan Wan, Shōsōin, and Carter versions. In the riding posture, with scarves swirling around them, they display bold facial features and proportions that may be described as short and stocky. They bring to mind Yan Liben's *Emperors Scroll* in the Museum of Fine Arts, Boston.[4] Yan Liben, of course, was the younger brother of Yan Lide.

Can there be a connection between Yan Wan's mirror, and by extension the Carter version, with the artist/craftsman Yan Lide? This is certainly an intriguing question, especially as one must recall that during Yan Wan's lifetime the development of the Tang mirror was moving at a fast pace. This was the height of the animal-and-grape type, the opulence of which easily surpassed the mirror in Yan

Wan's possession. Why did she hold onto a mirror, which was obviously already an anachronism, unless there was a special reason? Given her royal connection, she could have had a number of highly polished and elegant mirrors during her lifetime, without having to lavish attention on this simple and humble object. We may speculate that this mirror could have been designed by her father during his prime and then given to her as a token of affection; and years later, it became a fitting tribute to her remembrance.

A seventh century date for the inception of this type is likely, certainly long before Yan Wan's death, and probably during the prime of Yan Lide's career (ca. 620s). This typology with four riding immortals possibly can be placed in the second quarter of that century. A decade or so later, substitution of the mountain motif for two of the figures took place, resulting in a more stabilized design.

1. An excavated example from Xi'an, essentially identical to the Carter mirror, can be seen in Shaanxi 1959, pl. 119.

2. See Hubei/Yunxian 1987, 30–42, especially pp. 36ff, fig. 14.

3. See Shōsōin 1994, vol. 7, 227.

4. See Siren 1956, vol. III, pls. 72–75, especially pl. 75.

63.

狩獵菱花鏡
Octafoil Mirror with Hunters and Prey
Tang (618–907), late 7th–early 8th century
Weight: 899 g
Diameter: 19.9 cm
Rim height: 0.75 cm
Overall height (knob): 1 cm
1995.365

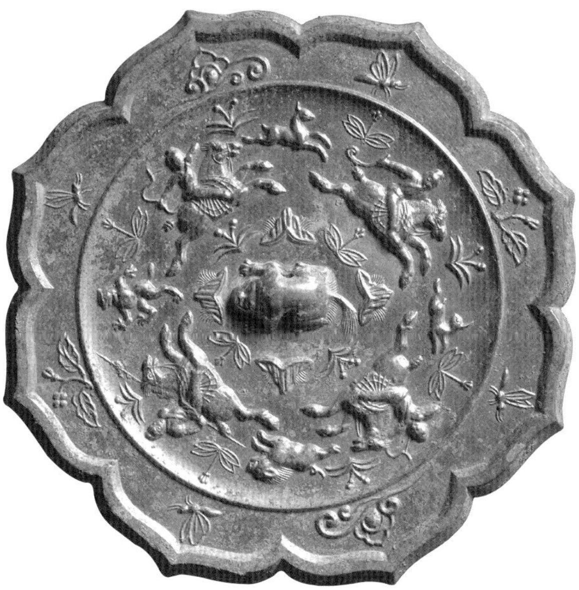

Corrosion has covered this mirror's surface with an uneven layer of green patination. Otherwise, this medium-sized mirror is strong in impact and dynamic in movement. The border arrangement is typical, with eight compartments for four types of decoration arranged in pairs across the same axis. Other than the cloud scrolls, which have been radically transformed so as to be almost unrecognizable, the rest are fine portrayals of insects and plants—a typical arrangement for Tang mirrors, perhaps of the eighth century.[1]

The chief focus, of course, lies in the inner circle, which has gained in size and dimension. At its center is a crouching beast. Around it, in a centrifugal direction, four mountains in diminutive scale set forth the cardinal directions, interspersed with four plants. Action is most apparent in the sequence of the hunt, with hunters on horses either urging their mount by means of whip, shooting with bow and arrow at the prey, or carrying a long lance for the attack. The prey includes bear, deer, rabbit, and boar—their size and pose being sufficiently varied. The insistence on *horror vacui* is apparent; wherever the likelihood of a vacant space exists, the mirrorsmith has inserted a plant to fill the void. The result is a fast-paced whirling motion that never ends.[2]

1. See Xu D 1994, 320–21.

2. For a comparable example found in Xi'an, see Shaanxi 1959, pl. 116.

64.

雙鳳雙獸菱花鏡
Octafoil Mirror with Paired Phoenixes and Animals
Tang (618–907), late 7th–early 8th century
Weight: 493 g
Diameter: 12.3 cm
Rim height: 0.8 cm
Overall height (knob): 1 cm
1995.345

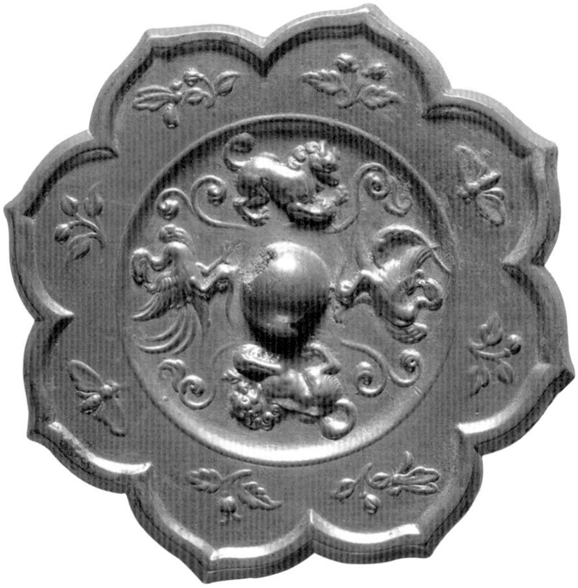

Being a small mirror with *linghua* configuration has its perils. It can be quite unwieldy, as the arched petals appear heavier in shape and in weight than the full interior circle. Otherwise, this mirror is well cast and well finished.

Typical of the Tang mirrors, each of the eight petals contains insect and floral motifs. The circle within has a hemispherical knob, around which phoenixes and *pixie* alternate in a revolving sequence. This arrangement, which emphasized the concentric circular conception, must have preceded the later, set viewing angle with the pairs of phoenixes in horizontal alignment and zoomorphic or floral decoration along the vertical axis.

This category of mirrors has been found in a site in Xi'an (burial date: 707)[1] and also in Yanshi, Henan province (burial date: 706).[2] It is suspected that the beginning of this typology could go back further, perhaps to the last quarter of the seventh century and continuing, unabated, to the eighth century.

1. See Shaanxi 1959, no. 132.

2. See Yanshi 1986, 429–62, pl. 5-(3), fig. 8.

65.

雙鳳雙獸葵花鏡
Lobed Mirror with Paired Phoenixes and Auspicious Animals
Tang (618–907), 8th century
Weight: 2,340 g
Diameter: 27.6 cm
Rim height: 0.8 cm
Overall height (knob): 1.2 cm
1995.350

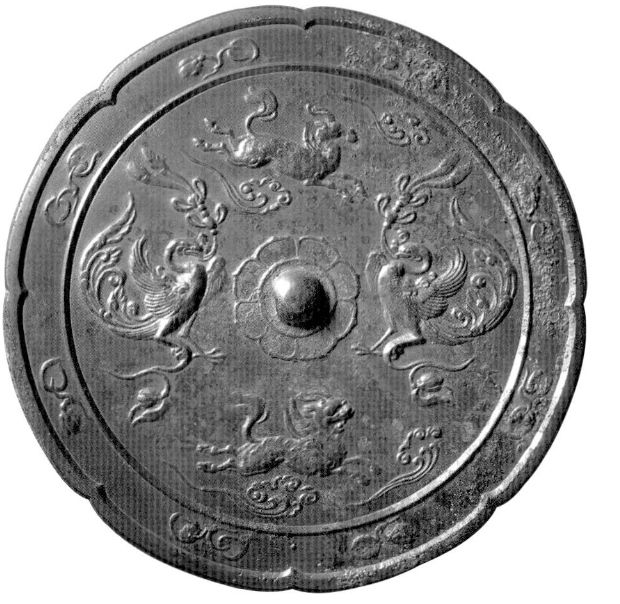

In a large mirror such as this one, the *kuihua* shape has now gained notable strength. The knob assumes the central core of a lotus blossom and is surrounded by eight petals. Beyond this central core, the inner circle gains in dimension, and the periphery, as defined by the raised ring and the lobed contour, is little more than a border, barely enough to contain eight plant sprays, compressed to fit the narrowed space. The most critical transformation, in relation to mirror no. 64, consists of the pair of confronting phoenixes, which sets the vertical axis in between, thus stabilizing the viewing angle. At the same time, they manifest the typical pose with their heads thrown back, chests projecting forward, wings spread out, and long tails a marvelous display of rich plumage. Seemingly about to venture forth, these birds plant one foot solidly on the ground, while lifting the other. From their beaks, a pair of tassels with two knots curve upward, a motif observed in mirror no. 61.

Uplifting as these divine birds are with their spreading wings and long sweeping tails, one is equally drawn to the energetic pair of galloping, fabulous animals which appear to leap over, or perhaps float above, a trailing cloud. The animal at the top resembles a deer but with a lone antler. That at the bottom appears to be a *qilin*, since unlike the *pixie* it has hoofs. On the other hand, *qilin* should be a unicorn, which is not the case here. Given the longstanding iconographic confusion, one may wonder if this pair, so carefully differentiated, could be the Tang interpretation of *tianlu* and *pixie*, guardian animals seen in Han or Six Dynasties tombs for eminent individuals. *Tianlu* (heavenly munificence) was known to be a unicorn, whereas *pixie* was typically endowed with two horns. The placement of these two beasts enforces the vertical viewpoint, especially as they are portrayed in a parallel way, and facing the same direction. This obviously stands as a rejection of the revolving images which, though typical of earlier mirrors, still managed to survive into later phases.[1]

By and large, this is a mirror of fine proportion, with a carefully coordinated structure and a convincing distribution of focal points. The *kuihua* shape here embraces the classic division between the interior circle and the border, but the former expands as the latter retreats to a narrow ring. Within this spacious interior, the finely wrought images in low relief and with incisive contour make the mirror all the more exquisite. Unfortunately, its condition is less than desirable, as redeposited copper patches interfere with our enjoyment of viewing.

1. Compare this with mirror no. 64 and a very impressive example in the National Palace Museum (Taipei 1986, pl. 106). In both, the revolving movement remains intact.

66a.

雙鳳花枝葵花鏡
Lobed Mirror with Paired Phoenixes and Floral Displays
Tang (618–907), 8th century
Weight: 692 g
Diameter: 14.8 cm
Rim height: 0.55 cm
Overall height (knob): 1.1 cm
1995.348

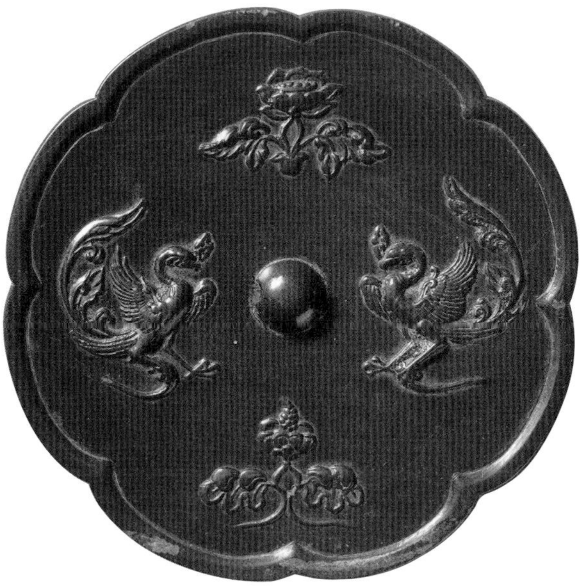

66b.

雙鳳花枝葵花鏡
Lobed Mirror with Paired Phoenixes and Floral Displays
Tang (618–907), 8th century
Weight: 676 g
Diameter: 15.6 cm
Rim height: 0.6 cm
Overall height (knob, including corrosion): 0.9 cm
1995.349

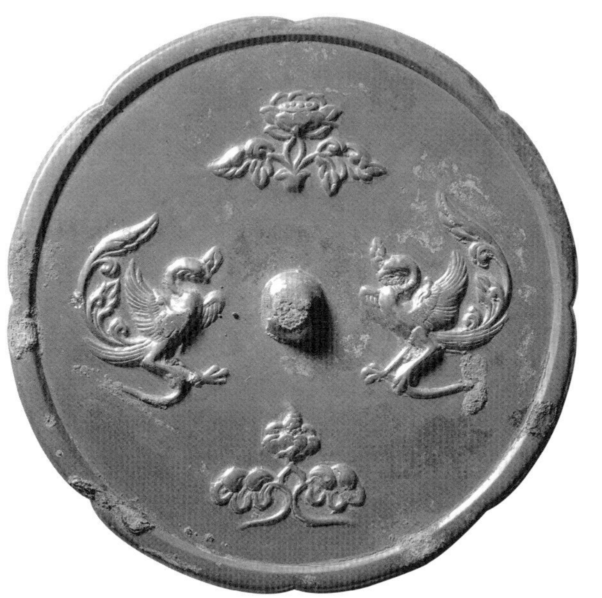

Of these two mirrors of nearly identical design, there is but one glaring contrast. Mirror no. 66a is an example of the post-burial condition known as *heiqigu;* its lustrous black surface is a clear contrast to the silvery tone of *shuiyinqin* in mirror no. 66b. Otherwise, both pieces are of the *kuihua* variety and display as a characteristic feature of that type a unified field within its enclosure. A pair of phoenixes is placed on opposite sides of the central knob; they help to set the vertical axis in between. Two isolated plants, placed at the extremes of that axis, are the key motifs. At the top is a pot of lotus blossom, flanked on both sides by the plant's broad leaves. Down below is a grapevine, the twin roots of which come together, then split apart, and finally join at the base of the flower and the emerging fruits, not yet ready to hang down like mature bundles. Possibly, this is a pictorial rendition of the so-called *lianlizhi* 連理枝 (linked branches), symbolic of love and devotion.

67.

雙鳳花鳥葵花鏡

Lobed Mirror with Paired Phoenixes, a Nestling Bird, and a Lotus Blossom

Tang (618–907), 8th century
Weight: 942 g
Diameter: 19.1 cm
Rim height: 0.65 cm
Overall height (knob): 1.1 cm
1995.346

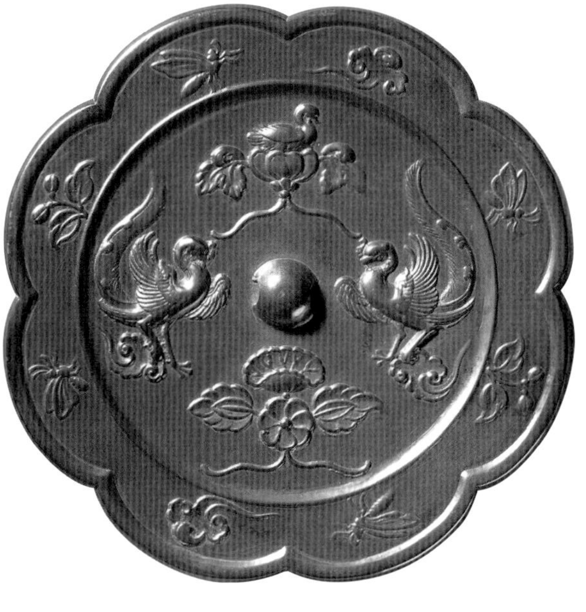

The division into dual compartments—a central circle and periphery—is clear and effective in this *kuihua* mirror. As in mirror nos. 66a and 66b, the pair of phoenixes once again set the orientation and help to steer the viewer to a vertical viewing position. The lotus is now placed below, with twin roots. It displays a blossom seen *en face*—a revolving circle composed of seven petals, two broad leaves extending to the sides, and a third one, visibly curled, at the top. Up above, a lone mandarin duck is seen resting on a floral throne, with grape leaves flanking either side. It, too, has twin roots, which spread and then rest in proximity to the beaks of the phoenixes. At the periphery, the eight compartments house bees, floating clouds, butterflies, and floral sprays, each poised across the axis opposite its own twin. In support of the animated presence of these subjects, the alloy is rightly lustrous, giving clear definition to each motif.

It may be timely to attempt the dating of this mirror series, a series known for the presence of paired phoenixes. Suggestion has been made that, among such mirrors, there are two instances in which twin phoenixes and floral motifs coexist and which are inscribed with the two characters of 千 (*qian*, or thousand) and 秋 (*qiu*, or autumns). *Qianqiu* (Thousand Autumns) recalls Emperor Ming's attempt to establish, on his own birthday, the Festival of a Thousand Autumns. This took place in 729 and lasted until 743. On that day, the emperor bestowed "mirrors of thousand autumns" on his consorts and officials as tokens of his affection.[1] If we agree that mirrors with such inscriptions were produced for this purpose, then a dating to the period between 729 and 743 would be appropriate. A plain mirror from Yangzhou, with the four-word inscription *qian qiu wan sui* 千秋萬歲 (Thousand Autumns and Myriad Years) is likely to be from the same era.[2]

This may be an oblique way to set a temporal reference to those mirrors with twin phoenixes and floral decors but *without* the *qianqiu* inscription. Such mirrors survive in great numbers, whereas mirrors of a "thousand autumns" are in a distinct minority. We can and should pose these questions: Did the court craftsmen initiate such a design at the imperial command and then afterward cause its spread throughout the country? Or did they modify and improve on an existing arrangement? In the first case, such mirror series would have to take the period of 729–43 as the point of departure, while necessarily lengthening the chronology to decades afterward. In the second scenario, these *qianqiu* mirrors would most likely stand at the medial point in the chronological scale that preceded 729 and decades after 743. Currently, datable samples of these non-*qianqiu* mirrors range between 745 and 778,[3] lending support to the first hypothesis. However, this can be easily overturned as future evidence comes to light. No matter, it is safe to consider that the production cycle of these mirrors fell within the period of the eighth century.

1. See Kong/Liu 1994, 122–23. In 743, Emperor Ming changed the Festival of a Thousand Autumns to that of Heavenly Lasting (*tianchang* 天長). Thus the *qianqiu* usage was limited to a fourteen-year period between 729 and 743. See also the explanatory text for mirror no. 72.

2. Taipei 1986, pl. 114. Another is published in Kong 1992, 580. These mirrors differ from the Carter series in that the upper and lower motifs along the vertical axis exhibit floral images together with square plaques in which the characters of *qian* and *qiu* would be inserted.

3. See Yen 1989, 336, chart 9, with a sequence of mirrors dated by means of burial sites from 745–78.

68.

雙鳳花鳥菱花鏡

Octafoil Mirror with Paired Phoenixes, Birds, and Flowers

Tang (618–907), late 8th century
Weight: 2,740 g
Diameter: 28.8 cm
Rim height: 0.95 cm
Overall height (knob): 1.6 cm
1995.366
(Plate 6)

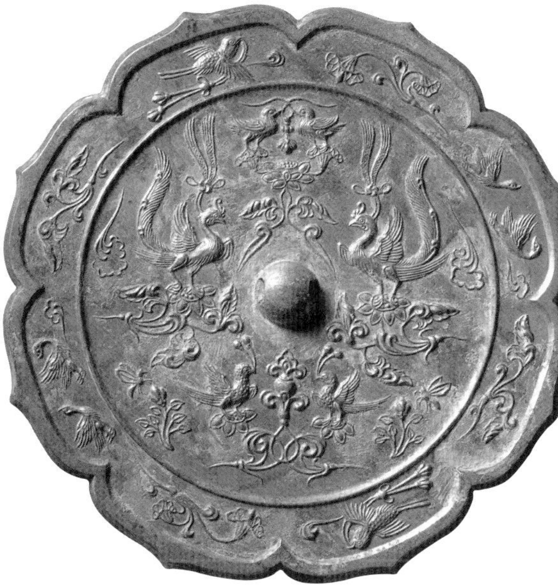

This is the largest of the Tang mirrors in the Carter collection. Richly decorated, the mirror is of the *linghua* variety, with its spacious interior punctuated by a hemispherical knob and surrounded by a border with the typical arched petal design. It also retains the paired phoenixes in its pictorial conception.

As elegant and graceful as ever, the phoenixes are nonetheless reduced in size and placed not on the axis to the left and right of the knob, but lifted slightly above it. No longer the controlling agents, they are merely part of an elaborate scheme. They stand on lotus blossoms and display their sumptuous plumage and poise, their wings and tails turning upward. Parallel to this movement are the thin ribbon scarves which emerge from behind their heads and float unchecked in rising curves. All these elements diminish the sense of gravity present in mirror nos. 66a–b and 67, and instead add a light airiness throughout.

In the vertical axis, a pair of mandarin ducks, symbol of conjugal fidelity, stand atop the leaves that flank an open lotus flower whose branch bifurcates and ends as trailing roots. Below, standing on two lotus blossoms, are another animated pair of birds sometimes identified as parakeets. Between them is a symbol that resembles a Buddhist *trisula*. All around is the play of swirling curves, either in the trailing clouds or in leaves or vines. Even the small insects and the sprays of flowers, seen at the lower end of the mirror's inner circle, participate in the endless movements up and down.

At the border, the eight compartments defined by the contour of the *linghua* contain four pairs of motifs, with identical ones at the dual ends of the same axis. The most notable of them depicts a parakeet in flight, its beak carrying a pair of tassels that float backward. The other avian motif is a flying pair of wild geese, the lead one turning its head to acknowledge its mate. Between them is a spray of lotus leaves as well as one that resembles grape leaves, neither of which carry any flower or fruit.

Given the elaborate sequence of images and the rich motifs, this may be described as the summation of the paired phoenix mirrors type—and a latecomer.[1] A late eighth century date is thus likely.

1. See Chen P 1987, pl. 72, which by the measurements supplied is slightly larger than the Carter example.

69.

雙鳳上圓下方葵花鏡

Lobed Mirror with Heaven and Earth between the Paired Phoenixes

Tang (618–907), mid or late 8th century
Weight: 1,533 g
Diameter: 22.5 cm
Rim height: 0.7 cm
Overall height (knob): 1.1 cm
1995.351

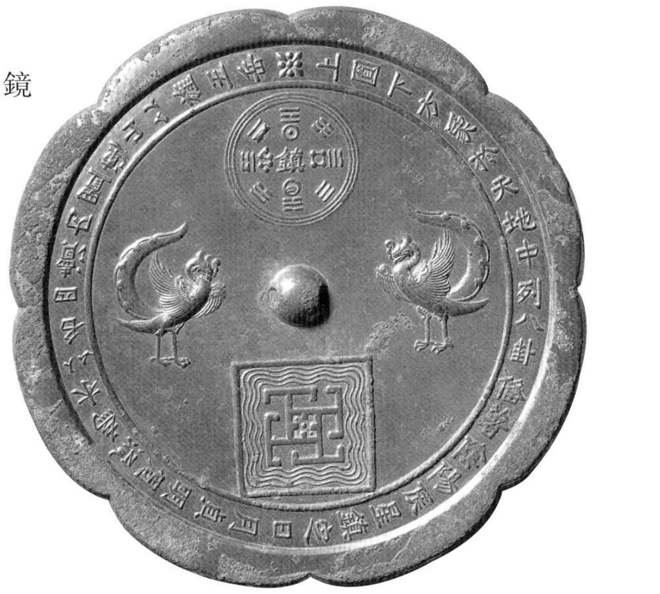

上圓下方，象于天地。
中列八卦，備著陰陽。
辰星鎮定，日月貞明。
周流為水，以名四瀆。
內置連山，以旌王岳。

Circle above and square below:
These are symbolic of heaven and earth.
In the middle are the Eight Trigrams,
Making clear [the forces of] yin *and* yang.
Stars and constellations are well aligned;
The sun and the moon are virtuous and bright.
Circulating around [the earth] is water,
Giving identity to the Four Great Rivers.
Within are the mountain ranges,
To signify the lordly heights.

This mirror's inscription helps to clarify the meaning of the diagrams. Heaven is above in the circle with the eight trigrams and sun and moon, and earth below in the square where the oceans and rivers surround the continents and mountains, depicted as four interlocking T's. This combination articulates the cosmology of Tang China, just as the TLV mirror presented that of the Han Chinese. Its owner did not merely see himself reflected in the mirror, but also within the whole universe.

Despite its popularity, no example of this type of mirror has been found in any of the excavated sites, nor with a date attached.[1] Setting its chronological framework thus will take some work. This type was obviously new when the twin phoenixes had already become a prominent motif; consequently, its maker felt inclined to retain that element even though there was no intrinsic reason to do so. The long inscription, too, was a revival after a period of prolonged neglect. Because of these considerations, we propose a middle or late eighth century date for this typology.

1. For a comparative example, see Lüshun 1997, 129, which bears the same inscription. Many more examples exist.

70.

真子飛霜葵花鏡

Lobed Mirror with a Tortoise Knob, a Musician, and a Phoenix

Tang (618–907), late 8th century

Weight: 1,394 g

Diameter: 21.6 cm

Rim height: 0.6 cm

Overall height (knob): 1.2 cm

1995.369

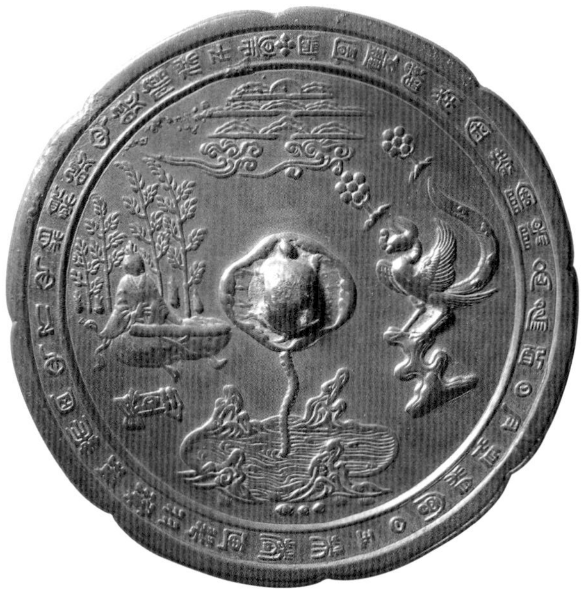

Unlike previous examples, this mirror displays only one phoenix—a virtual vestigial remain. It is pictured perhaps only as a force of habit, and also as a device to fill in vacant space. Without its mate, this phoenix is singularly out of place in the narrative scheme. It is a reminder of a long tradition that has run its full course.

To understand this mirror and its decoration, begin from the lower end of the vertical axis. Note the small pond surrounded by four "sacred mountains." From the pond, a tall vertical stalk rises and terminates in the broad leaf of a lotus, which is seen *en face*. That leaf becomes a platform on which lies a tortoise in the form of the mirror's knob. Further up, trailing clouds drift by, with layers of hills behind, and a sun (or is it a moon?) emerges from the clouds. To the right is a rocky promontory upon which stands the phoenix in a typical pose of thrown-back neck, half-extended wings, and upward tail. Directly above, two trees, perhaps prunus, are planted at an angle to the assumed line of the horizon. To the left of the tortoise knob is a hermit who plays the *qin*. In front of him is a small stool with writing implements, behind him a bamboo grove.

Thus we have a pictorial rendition of a scene, with images of mythological import all around its core. The inscription reads:

鳳凰雙鏡南金裝。陰陽各為配，日月恒相會。白玉芙蓉匣，翠羽瓊瑤帶。同心人，心相親，照心照膽保千春。

A mirror with paired phoenixes
Is cast from metals from the South.
The yin *and* yang *are made to match;*
The sun and moon meet with constancy.
A white jade box with peony decor
Is mated with a lustrous tassel of kingfisher feathers.
Those of the same heartbeat,
Are drawn close.
Whatever the mirror will reveal: be the hearts or gall;
It will surely protect you for a thousand springs.

Mirrors like this one are often known under the inscribed title of *Zhenzi fei shuang* 真子飛霜—"Zhenzi Causes the Frost to Descend," certainly an enigma. Some of the mirrors carry such a title inscription, though it may be justified to designate the whole series by the same title. Is Zhenzi (he who has attained the true nature) the musician in the scene? Then what does *fei shuang* (spray the frost) mean? Could it refer to the title of a composition of *qin* music? Unfortunately, all inquiries have so far led to a dead end.

In rare instances, Hou Jinzhi 候瑾之 appears in cartouche above the tortoise.[1] If we were to put Hou Jinzhi and Zhenzi together, then one can assume that Hou Jin, a Han dynasty hermit, took on the appellation of Zhenzi, whose playing of the *qin* caused the frost to form or the snowflakes to fly. But as much as Hou Jin was admired as a hermit and writer, no one has ever suggested that he was a musician. Perhaps he became so in the eyes of Tang Chinese. Whatever the mystery, this particular mirror type is widespread during the Tang period—from Xi'an to Yangzhou and from Zhejiang to Jilin.[2]

1. Umehara 1984b, pl. 63a. Also, the same inscription appears in Liu T 1935, *juan* 16, 81b. According to the *Hou han shu*, Master Hou was a native of Dunhuang and was admired for his eremitic aspirations and lifestyle. However, in the biographical account of *Hou han shu*, he was not at all known to be a musician, but remembered as a writer of essays. See Zhang Y et al. 1967, vol. 5, 3780.

2. Excavated examples cover a fairly large territory. See Kong 1992, 629–34, p. 634 shows a later piece from Xi'an; Wang SL 1987, no. 119, with inscription of "Zhenzi fei shuang"; Zhou/Zhou 1979, 54–55, pl. 15. Another piece was unearthed at Hengyang, Hunan; see Tang 1992, 1054; Wu S 1993, pl. 69; Zhang 1990, app. no. 1. From Ninghai, Zhejiang, came another piece with the proper title; Teng/Shi 1993, 62. The authors observed that the trees represented above the phoenix are prunus, and that this type of mirror was used in a marriage ritual to symbolize the expected conjugal happiness. The reasoning here needs to be clarified, even though it seems self-evident to the authors.

71.

三樂鏡

Lobed Mirror of Three Delights

Tang (618–907), 8th century

Weight: 368 g

Diameter: 12.7 cm

Rim height: 0.55 cm

Overall height (knob): 0.9 cm

1995.363

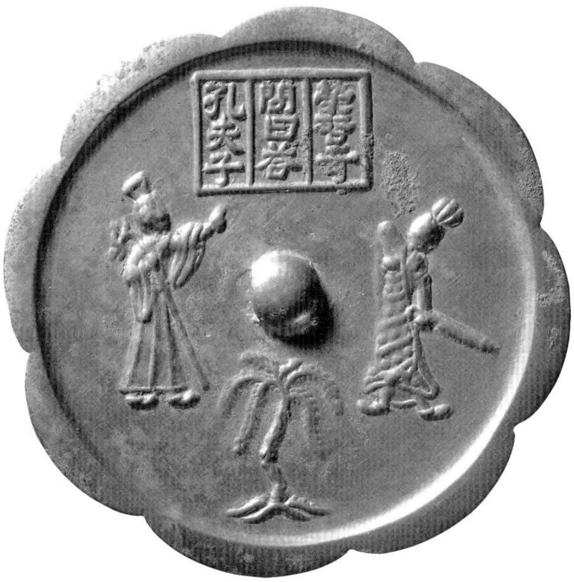

A familiar type, the popularity of this mirror enabled it to spread far and wide. As already noted in a number of cases, the pictorial scene at the back of this mirror has been softened through polishing or handling. A finer example may be seen in the National Palace Museum.[1] Typically, the framed inscription is placed at the top, making clear the identity of the two protagonists: 榮啟期問曰答孔夫子 (Rong Qiqi was questioned by, and replied to, Confucius).

The subject of this mirror involves two august figures in history.[2] Confucius here assumed a secondary role. He was the seeker of truth whereas Rong Qiqi played the man of attainment, already wise and sagacious. The circumstances, which are interpreted through the eye and words of Liezi 列子, may be fictitious. The "Three Delights," a title often given to this mirror type, stems from the hermit's declaration that, first of all, he was born a human being with intelligence above all other forms of existences; secondly, he was born a man, at the time when society valued man more than woman; and thirdly, he enjoyed longevity, having lived for ninety-some years.

The slender proportion of the two figures, with Confucius at the left and Rong Qiqi, clad in a deerskin garment, at the right, is congruent with the earlier phase of the Tang period.[3] They are contrary to the figure style of the High Tang, where a full body and ample proportions were the rule. Consequently, we may suspect that the prototype of this mirror may have already existed during the early eighth century.

1. See Taipei 1986, pl. 130. Also, compare it to archaeological examples, e.g., Shaanxi 1959, pl. 117; and Luoyang 1988, pls. 87, 89.

2. The encounter was narrated in *Lieze* in the "Tianrui" chapter; see Liezi 1929, 4b.

3. For comparison with procession figures on the wall paintings in Princess Yongtai's tomb, see Beijing 1963, color pls. 1, 2; the latter reproduces a lady attendant with round head and slender body, quite similar to the figures in this mirror.

72.

盤龍鏡

Mirror with a Coiling Dragon

Tang (618–907), 8th century

Weight: 329 g

Diameter: 10.2 cm

Rim height: 0.8 cm

Overall height (knob): 1.1 cm

1995.367

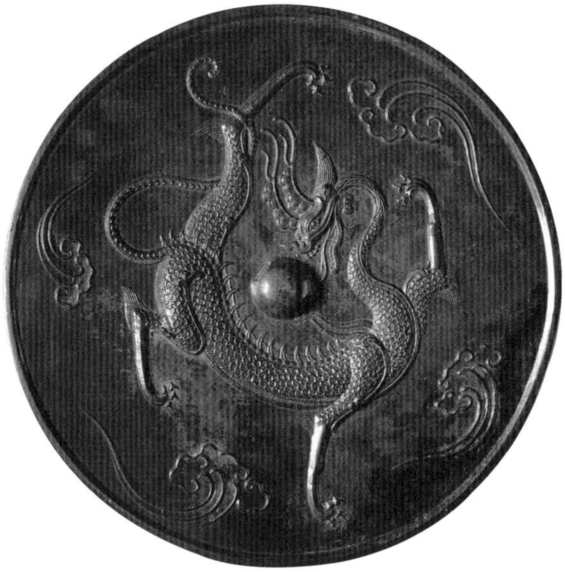

Against a plain backdrop and a unified field made somber by dark-toned *heiqigu*, this high-tin mirror displays a single scaly dragon intertwined around the knob, turning it into the proverbial pearl. Around the dragon are four trailing clouds. The divine creature twists and turns with compressed vitality and dynamism—a simple, effective design.

The dragon of the Tang dynasty is unique among others. Its eyes, brows, and nose are concentrated in one area of its distinctly small, elongated, horned head. Wide-open jaws allow its tongue to touch the pearl of the knob. Linked with the head is a neck much longer and thinner than in any other dragon portrayals. Its scaly torso soon becomes fuller, with a row of spikes on its back. The strong legs are held wide apart, seemingly invincible in their forward flight, and end with three tensely held claws. Its tail once again becomes long and thin, twisting around a hindleg before curling upward.[1]

Comparable examples are plentiful,[2] with Xi'an and Luoyang tombs yielding similar specimens.[3] Of great importance are two datable examples: a mirror fragment bearing the dragon design from the 741 tomb of Zhang Jiuling (673–740), a famous official; and another mirror from Yanshi, Henan, with a *terminus ante quem* of 738.[4] Other mid eighth century examples have also been found.[5]

While this confirms an eighth century chronology for the type, we must remember that these dragon mirrors may have been in their owners' possession for decades. Thus, the beginning of this type could be as early as the 720s, and it may have continued unabated for some time.

It so happens that there were also *qianqiu* (Thousand Autumns) mirrors with the dragon image.[6] Since one type of *qianqiu* mirror contains paired phoenixes as its main imagery, then the dragon could be considered even more appropriate for mirrors with a connection to Emperor Ming of the Tang dynasty. If so, then the original production of this series would have taken place between 729 and 743. This would allow a retiring Zhang Jiuling to carry his courtly mirror back to Guangdong and then, at death, have it buried by his side.

While the invention of this type seems to have taken place at court, it soon spread beyond its confine. The awareness of imperial favor and sponsorship must have created waves of intense interest, leading to numerous imitations and homage. These productions soon outnumbered those of the court. Thanks to the excellent level of craftsmanship during the High Tang, the result of these imitations is quite impressive.

1. For reference, see Xu Y 1986, 58–63. The author's main interest is dragon portrayal in jade objects, but he also refers to similar images on mirrors.

2. See Kong 1992, 646–51.

3. For Xi'an examples, see Shaanxi 1959, pls. 146–47. For Luoyang finds, see Luoyang 1988, pl. 92.

4. For the excavation report of Zhang Jiuling's tomb, see Guangdong/Huanan 1961, 45–51, fig. 16. For the Yanshi mirror, see Yanshi 1986, 429–62, pl. 5-(5); Kong 1992, 651.

5. See Yen 1989, 338, chart *chen*, under the title of *panlong jing* (dragon mirrors).

6. See explanatory note under mirror no. 67. For an example, see Shaanxi 1959, pl. 146; Lüshun 1997, pl. 118.

73.

蟠龍紋鏡
Mirror with Interlacing Dragons
Tang (618–907)
Weight: 621 g
Diameter: 16.9 cm
Rim height: 0.6 cm
Overall height (knob): 0.6 cm
1995.364

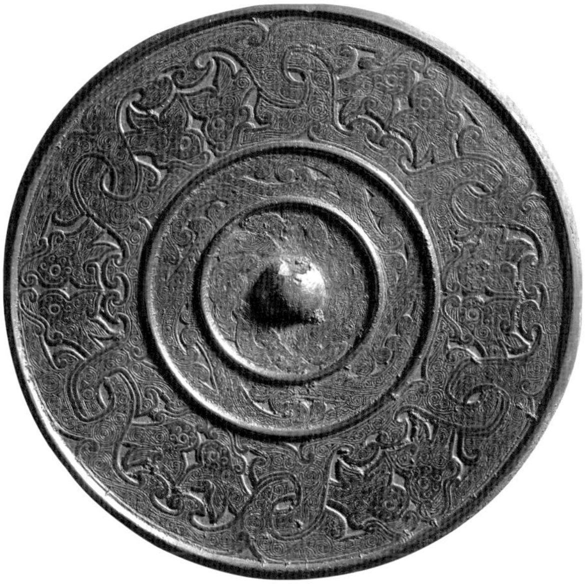

Amid Tang mirrors, this typology stands as an oddity. Were it not for archaeological finds,[1] one might easily mistake this mirror as coming from an earlier era—the late Zhou, for example. The broad band of interlaced patterns is characteristic of earlier types, and its very flatness is contrary to the Tang period's volumetric tendency. Only the high ridge of the concentric rings and the glossy luster betray the later date.

The most intriguing aspect of this object is that, prior to the Tang, there are no known mirrors in exactly the same vein, even though the decoration was directly taken from ancient bronze vessels. This being the case, it can be assumed that this is a product of a society that, from time to time, felt compelled to look toward the ancients, without attempting to re-enact the same. Instead of mimicry, this mirror has been created more in the vein of homage.

1. See Luoyang 1988, pl. 93; Farley 1940, 72–94.

74.

雙獅亞形鏡
Cruciform Mirror with Two Lions
Tang (618–906), late 8th–early 9th century
Weight: 214.5 g
Dimensions: 10.4 x 10.5 cm
Rim height: 0.35 cm
Overall height (knob): 0.7 cm
1995.373

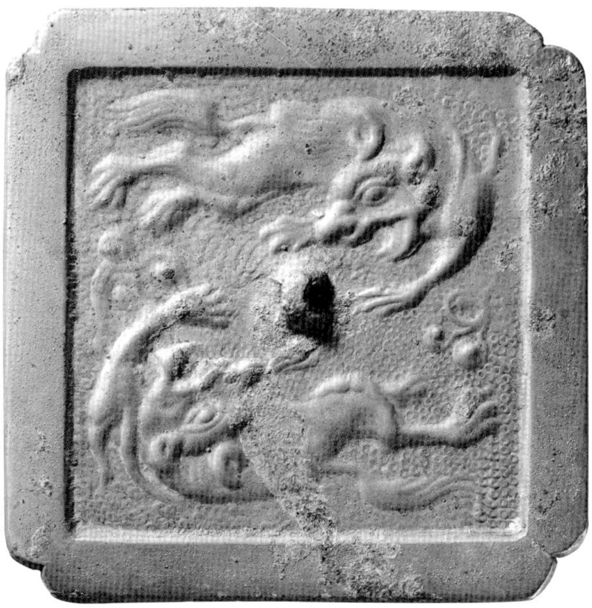

Archaeological sites have yielded a number of square mirrors with indented corners, vaguely resembling a cruciform shape. In this example, the plain border encloses a square wherein the knob assumes the form of a slender arc, which widens slightly at the base. Across its axis, on both sides, are two inversely posed lions with twisting bodies, gaping mouths, and strong forward claws. Still, their air is less threatening than whimsical. The ground is enlivened by pearl-like shapes, obviously done to simulate a *repoussoir* technique. The roughness of the casting contributes to an uneven display, as some elements are clearer than others. A limited amount of corrosion has invaded the back of the mirror. The silvery luster of its surface still links it to the last years of the Tang era or to the early Song period.

As a matter of fact, the earliest known example comes from a Tang burial site in Yangzhou, about which little is published.[1] From Luoyang, a Song burial site at Yuejia Village also yields an example.[2] Compared to this Carter mirror, the Song example is just a shade smaller (9.04 x 9.04 cm), although its weight is significantly less (only 115.5 g).

The inverse poses of the two zoomorphs are not unique, but rather part of a larger pattern. We may compare this to other Tang mirrors with avian motifs similarly arranged; a pertinent example may be the *kuihua* mirror found in a 745 tomb near Xi'an.[3] Other examples are found in a burial that contained the remains of Zheng Xun and his wife, who died in 768 and 776, respectively.[4] It appears likely that the avian specimens were produced in the second or third quarter of the eighth century, and the lion version came after.

1. See Zhou/Zhou 1979, 53ff, fig. 1.

2. For comparative purposes see Luoyang 1988, no. 188. For measurements and weight see ibid., "Explanatory Notes," 23.

3. See Shaanxi 1959, pl. 141.

4. See Yanshi 1996, 1–35, figs. 20-(3), 20-(4).

75.

卍字鏡

Quadrilobed Mirror with 卍 Symbol

Tang (618–907), 8th–9th century
Weight: 399 g
Dimensions: 14.1 x 14.3 cm
Rim height: 0.4 cm
Overall height (knob): 0.9 cm
1995.362

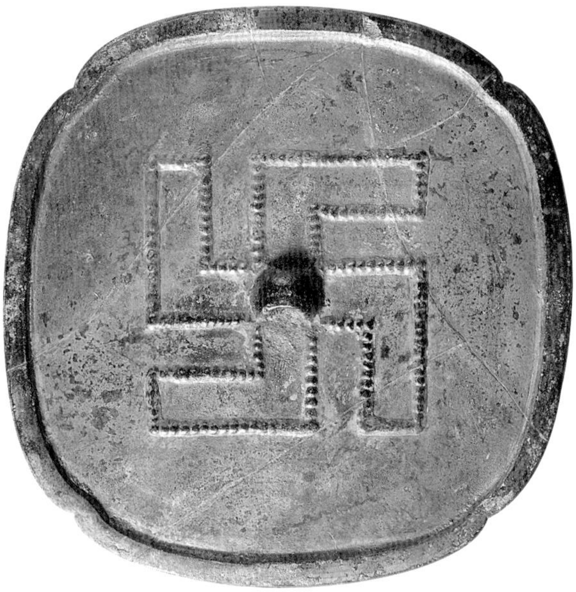

The 卍 emblem indicates this is a Buddhist mirror, and thus would have been buried with the remains of someone who adhered to Buddhist teachings. It is a far cry from more elaborate and sumptuous types, such as the animal-and-grape mirror. Here the mirror body is thin and the casting barely acceptable. But the simple state of this mirror also makes its message: the spiritual is far more important than the material remains.

Two dated examples have come to light in recent years. The earlier version is from a tomb in Yiyang, Hunan, with a date of 763.[1] Another, more elaborate example—containing four additional characters among the strokes that form the 卍—is from a tomb site in Shanxian, Henan, dated 838.[2] For this type of mirror, one can say simply that the second or third quarter of the eighth century may well be the time of its inception. That it survived into the ninth century and beyond is easily testified to by another mirror in the Carter collection (see no. 76).

1. See Yiyang 1981, 313–18, and fig. 2. Another example also came from the same region, and may be dated around the same time.

2. See Kong 1992, 666.

Phase IV

Liao, Song, and Jin Dynasties 916–1279

76.

卍字鏡

Mirror with 卍 Symbol

Liao (916–1125), Taiping reign (1020–30)
Weight: 318 g
Dimensions: 12.5 x 12.5 cm
Rim height: 0.45 cm
Overall height (knob): 0.9 cm
1995.368

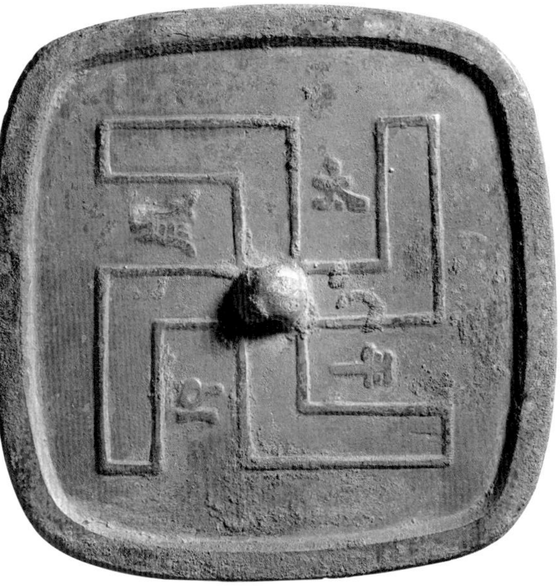

We have already noted that mirrors with the 卍 symbol have been known since the middle or late Tang period (see no. 75).[1] The act of interspersing smaller Chinese characters among the symbol's arms was already a *fait accompli* by the second decade of the ninth century.[2] It only remains to be said that while the usual wording is *yun shou zhi jing* 永壽之鏡 (mirror of eternal life), this mirror's inscription reads, *tai ping wan sui* 太平萬歲 (Long Live the Taiping Era). Taiping refers to the reign of Emperor Shengzong of the Liao dynasty (r. 1020–31). As is well known, Liao artists and craftsmen inherited the Tang traditions before evolving their own interpretation. This is one such example.

1. See Kong 1992, 665; this tomb was dated 838. A circular Liao mirror with the same Buddhist symbol was also found in Pingquan, Hebei; see Hebei 1996, pl. 66.

2. See Kong 1992, 666.

77.

雙鳳花枝方鏡

Square Mirror with Two Phoenixes and Floral Sprays

Northern Song (960–1127)
Weight: 723 g
Dimensions: 16.2 x 16.2 cm
Rim height: 0.5 cm
Overall height (knob): 0.8 cm
1995.374
(Plate 7)

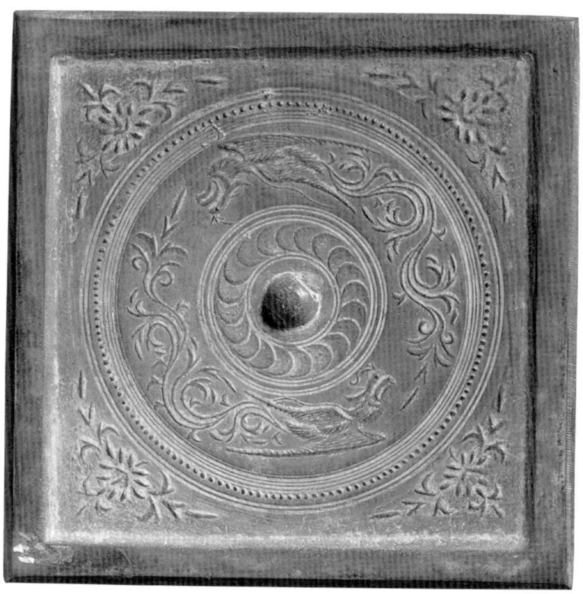

Square in shape with a circle motif within, this mirror is the very opposite of the TLV mirror (nos. 27–29) of the Han dynasty. It is as if heaven and earth have changed places; but as far as the mirror is concerned, it is none the worse. Easy on the eyes, its appeal is subtle.

The plain square border is reminiscent of late Tang mirrors (see no. 74), but in other respects this Song mirror is the very antithesis of Tang heritage and style. In contrast to the thickened disc, high relief, and sumptuous conception that characterize so many of the Tang mirrors in the Carter collection (e.g., nos. 55, 65, and 68), this mirror subsides into understatement and reserve. Gone are the clear-cut layout and tectonic structure, in favor of gradualism in pace and tempo. This is clearly illustrated in the formation of the outer circular ring. Instead of a strong ridge supplemented by one or two secondary ones, this ring is built up with five concentric circles moving from the inner register upward to tip the scale at the circle of beads; it then requires two additional concentric layers to prepare us for descent to the other side.

At the core, a ring of overlapping discs circles the knob, leaving essentially the crescent shapes. The principal band is sufficiently wide to accommodate a pair of phoenixes that chase each other. They glide forward gently and effortlessly, with wings extended backward and tails dissolving into a loose complex of undulating curves and curlicues. In the corner space between the outermost ring and the square border are four sets of floral motifs loosely formed into symmetrical images. Each has a chrysanthemum blossom at its center, with leaves extending to its sides and additional leaves at its apex.

Throughout the design, an even tempo prevails. The only tension found is focused on the knob and the adjacent ring of overlapping discs—a tension that releases itself gradually as the mirror expands its concentric circles and then eventually changes its shape into that of a square.

It is important to note how this particular design shares traits with decorated bricks of the Tang period. Perhaps the closest example is the square brick, the center of which is decorated with a circular double ring with a central lotus blossom. This brick design leaves four corners to be filled with vegetal motifs far simpler than those in this Carter mirror, but could have inspired the Song mirrorsmith to engage in further elaboration. For these brick designs we refer in particular to those found in the ruins of the Yuhua Palace, founded as early as 624, and later transformed into a Buddhist temple in the year 659.[1] The first indication of transference of this type of design to the mirror occurred during the Five Dynasties period (907–60), before going on to influence the Song mirrors.

1. See Lu 1978, 380–87, fig. 4. Yuhua Palace was where Emperor Taizong of the Tang dynasty honored the great pilgrim, Xuanzhang, by promulgating the "Preface to Holy Teaching 聖教序." It was in this same site that Xuanzhang translated sutras from Sanskrit to Chinese, sutras that he had brought back from India.

78.

亞字形纏枝六花鏡

Cruciform Mirror with Six Blossoms

Northern Song (960–1127)

Weight: 408 g

Dimensions: 16.5 x 16.5 cm

Rim height: 0.35 cm

Overall height (knob): 0.7 cm

1999.222

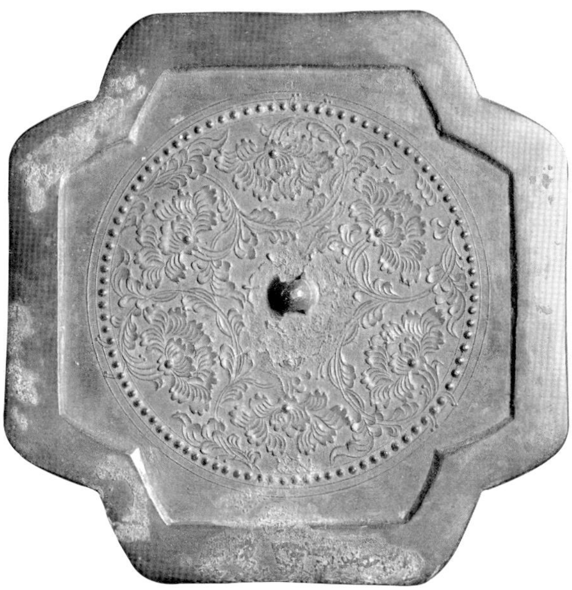

This mirror has a cruciform shape with a plain, broad border and a small knob, around which poor casting or damage has eradicated the surrounding petals. A fine tracery of branches and leaves encloses six blossoms whose petals open and turn to unveil a pistil with a small bead at the core. The branches also produce twin buds, several of which protrude beyond the beaded circular border to touch the edge of the outer circle—a lighthearted and whimsical touch. Nevertheless, the decorative pattern's intricacy calls attention to its tendency for release and involutions. The same decorative pattern is also known in silver objects and ceramics of this period. Here, however, the subtle change in levels of relief creates an undertow of flickering lights and surging waves.

From Luoyang, a Song tomb yields a similar cruciform mirror with floral decoration based on interlaced branches, vines, and blooming flowers.[1] A less mature example comes from Xi'an, where it was buried in a tomb dated to 1119.[2] This chronological framework thus places this type of mirror in the late Northern Song period, from which it also extended its network of distribution over the Liao and Jin territories.[3]

1. See Luoyang 1988, pls. 171, 174.

2. For a mirror of similar, if not identical design, see that found near Xi'an in a tomb dated 1119; Shaanxi 1959, pl. 167.

3. For a similar mirror, see Liu S 1997, pl. 107; for Jin counterparts, see Hebei 1996, pls. 361–63.

79.

雙鳳鏡
Mirror with a Pair of Phoenixes
Song (960–1279), 12th century
Weight: 3,760 g
Diameter: 35.1 cm
Rim height: 0.75 cm
Overall height (knob): 1.3 cm
1995.389

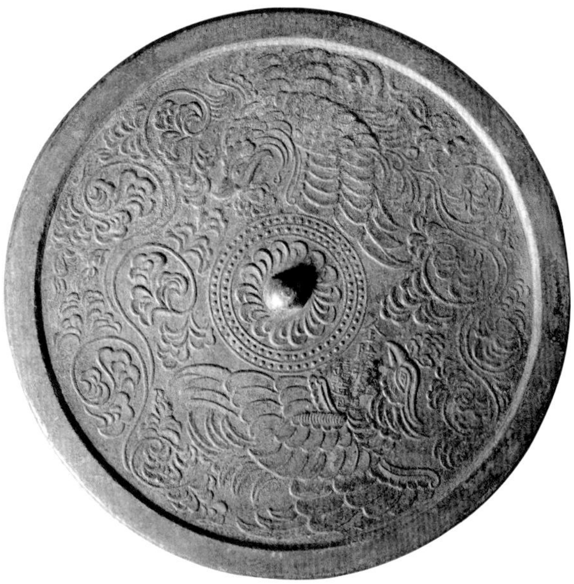

This mirror, the largest in the Carter collection, has a plain border and a circle of chrysanthemum petals surrounding a small knob, which in turn is encircled by three rings in between two strings of beads. Within the broad surface so created, a pair of phoenixes chase each other, their rich plumage turning into extended traceries more floral than feathery in appearance.

So far there appears to be no identical mirror known, either in published collections or in excavation reports. In type and design, it could perhaps be traced back to simpler mirrors with twin phoenixes, such as that found in Li He's tomb in Yanshi, Henan province, dated 843.[1] Li He's mirror, however, does not display the central chrysanthemum. This feature was first seen in a mirror of the Five Dynasties period,[2] without the twin phoenixes. The chrysanthemum motif then gradually spread to later mirrors (presumably Song) of varying types.

One can conclude, then, that this Carter mirror may indeed be a later type that combines the twin phoenix motif and the central chrysanthemum into a coherent design. Also of note is the chrysanthemum's petals, which adhere to a major and minor alternation, obviously a further modification of its Five Dynasties predecessor whose petals all took the same size and shape.

Lending support to this chronology are the data from our study of the mirror's alloy composition, which shows a moderate decrease in tin and a notable increase in the lead level, closely matching that of the tin.[3] Recent studies confirm that while the mirrors of early Song and Jin adhered to a high-tin, low-lead formula, later Song and Jin mirrors tended to observe the very opposite: a low-tin, high-lead composition. This mirror stands somewhere in between, in a transitional stage. It is thus possible to date this mirror to late Northern Song or early Southern Song.

Why must this mirror have a Song provenance and not one of Liao or Jin? From the inscription on the mirror, we learn that the mirrormaker, Zhang Zhigao 張子皋, came from historic Jinling 金陵, or Nanjing. At that time Nanjing, located south of the Yangzi River, was part of the Song territory.

1. For a line drawing of this mirror, see Xu D 1994, 299–342, fig. 6. For Song examples, see Zhou SR 1986b, pls. 153–55; Hebei 1996, pl. 185. For Liao examples, see Liu S 1997, pls. 53, 60, 130, 132, 175. For Jin examples, see Hebei 1996, pls. 327, 328; Kong 1992, 827.

2. See Zhou SR 1986b, pl. 137.

3. See this catalogue's Appendix, under CMA 1995.389.

80.

佽非斬蛟龍葵花鏡

Octafoil Mirror Featuring Ci Fei, the Dragon Slayer

Song (960–1279) or Jin (1115–1234)
Weight: 380.5 g
Diameter: 17.2 cm
Rim height: 0.45 cm
Overall height (knob): 1.1 cm
1995.376

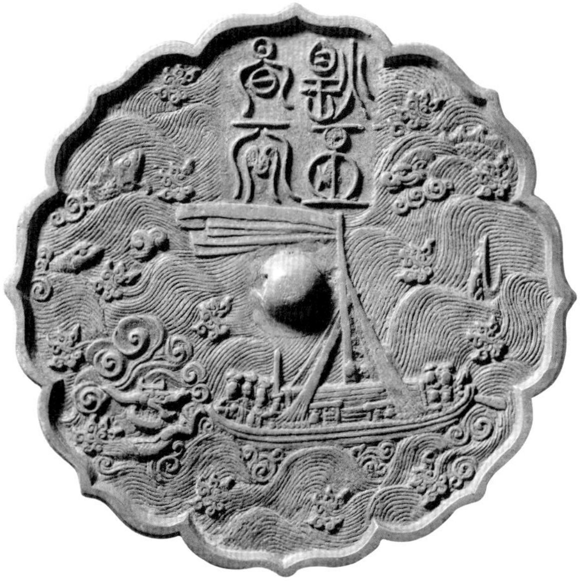

In this mirror a large ship sails through turbulent waters carrying a load of passengers, one of whom raises a sword to fend off a fiery dragon that is chasing the boat. Another partial apparition of a large leviathan is seen ahead of the ship. The scene is reminiscent of the story of Ci Fei as narrated in Liu An's *Huai'nan Zi*, a work of the Han dynasty.[1] A fearless hero, Ci Fei ventured forth and obtained a fine sword, and on his return journey was aboard a ship being attacked by two *jiao* dragons. Realizing that the ship was doomed, Ci Fei leaped into the water and used the sword to kill the creatures. The inscription on this mirror is in seal script above the craft. It reads, *huangpi changtian* 煌丕昌天 (Great and resplendent is the heaven), which obviously refers to Ci Fei as a savior figure and appeals to the grace of the heavens.[2] A similar mirror bears a different inscription, *tianxia anchang* 天下安昌 (Peace and prosperity under heaven), which alludes to the same message of a peaceful and prosperous future.

Incidentally, Ci Fei's story was the subject of painting as early as the Tang period. The great poet Li Bai 李白 (701–762) composed a poem in celebration of such a painting, which ends with:

感此壯古人，千秋若對面。

I'm moved by this brave man of the past.
Even after thousands of years,
I can see him in my mind's eye.[3]

Most of this type of narrative mirror are of northern provenance, which spread as far as the northeast and to Korea. Kong Xiangxing published a piece excavated from Qiange'erluoshi, Jilin; he also included one with an *ex post facto* engraved inscription that reads *Lintao fu lu shi si yan ji guan* 臨洮府錄事司驗記官, which indicates an inspection at Lintao Prefecture, Gansu province.[4] More important is the site of another find at Zhi'an, Sichuan; this same tomb also produced a Xuanhe coin, Xuanhe being the reign of Song Huizong (1119–26). Hunan, too, has yielded a Song specimen, important because of its southern linkage.[5] In short, this type of mirror may have emerged before the last years of the Northern Song period in China, and subsequently spread throughout the broad territories of northern China under the Jin as well as to Korea.

As typical of mirrors, its modest size made for easy mobility. Song mirrors can be found in Jin territory, just as Jin mirrors can be seen in the south. The mirror bearing the Lintao inspection mark points to a Jin provenance at a given time; however, it does not necessarily imply a Jin origin. It tells us that an official sanction was given to this particular mirror, which could have been either made locally or imported. Such official sanctions were due to a copper shortage, which led to restrictions in unofficial casting of bronze artifacts, mirrors included.

1. See Liu A 1940, *juan* 12, 454.
2. In the catalogue of the mirror collection at the National History Museum, Beijing, its author proposes an alternate reading, "亟" for "丕". See Yang 1994, 131, relative to pl. 37.
3. See Dong et al. 1962, vol. 8, 4485.
4. See Kong 1992, 848–49. This mirror, or a rubbing of it, must have been reprinted from Xu N 1914, *juan* 1, 25a.
5. See Zhou SR 1986b, pl. 170.

81.

牛郎織女菱花鏡
Octafoil Mirror with Legend of Herdboy and Weaver Maid
Song (960–1279)
Weight: 638 g
Diameter: 17.8 cm
Rim height: 0.7 cm
Overall height (knob): 1.1 cm
1995.383

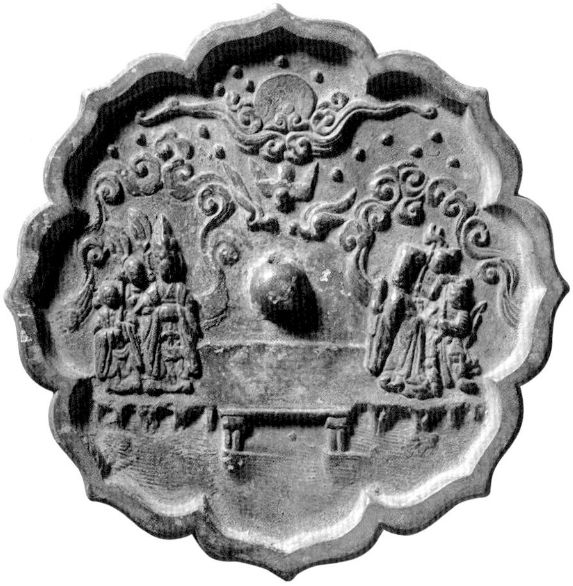

An old legend, already popular in antiquity, is depicted in this mirror. It tells the story of the celestial constellations identified as the Herdboy and the Weaver Maid, daughter of the Emperor-in-Heaven. Year after year, the lonely maid toiled at her weaving. Emperor-in-Heaven took pity on her and so arranged for her marriage to the Herdboy. Wallowing in marital bliss, she neglected her duties and thus her displeased father separated the couple. He allowed them to meet only once a year in the Milky Way when, on the night of the seventh day of the seventh month, magpies formed a bridge. On that evening, the Chinese could look upward to the skies and observe the starstruck lovers reunited.

The mirror shows the Herdboy at right and the Weaver Maid at left surrounded by their respective entourages. Clouds swirl about them as they stand at opposite sides of the bridge. In between, three magpies form a triad, announcing the impending reunion. The moon floats above the clouds.[1]

An identical mirror, unearthed from Hunan, has been dated to the Song dynasty. This is not surprising, since most of the narrative mirrors had their origins in Song China, where the mirrorsmiths avidly pursued this genre in response to clientele demand.[2]

1. For a comparable but finely cast example, see Rupert/Todd 1935, pl. XI.

2. See Zhou SR 1986b, pl. 165, which is quite similar in terms of layout and composition.

82.

渡海二仙菱花鏡
Octafoil Mirror with Two Immortals Crossing the Ocean
Song (960–1279)
Weight: 606 g
Diameter: 17.6 cm
Rim height: 0.6 cm
Overall height (knob): 1 cm
1995.378

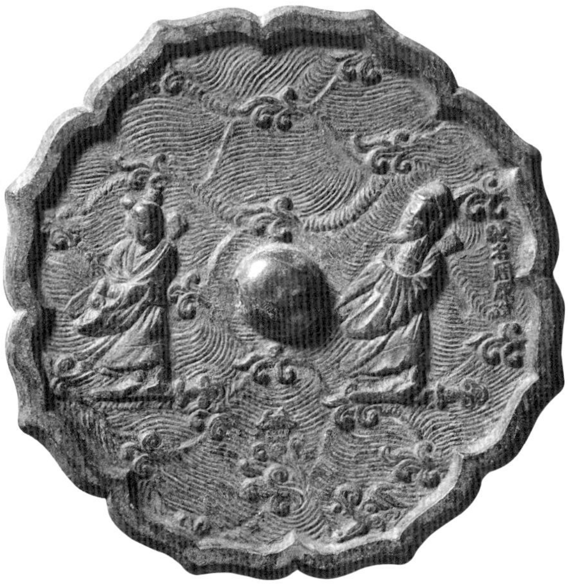

This mirror, as the inscription makes clear, is from the hand of Zhou Yuan. Apparently he enjoyed a degree of fame as a craftsman, thus the self-proclamation: 馳名周原造 (Made by the famous Zhou Yuan). Elsewhere, he was content to sign only "Zhou Yuan."[1] We assume that he did so early in his career.

On this *linghua* mirror's back, enclosed in a broad, eight-petaled contour, is a full view of palpitating waves. Two immortals float on them, each standing on a sword. Below the round knob is a small palace, perhaps home of one of the dragon kings. It has been suggested that the figures are none other than Zhongli Quan 鐘離權 and Lü Dongbin 呂洞賓, two well-known immortals from a group of eight. This is still subject to further confirmation, even though the figure at left, who wears his hair in two knots, conforms to Zhongli Quan's reputed appearance, and the figure at right could easily pass as Lü Dongbin. During the Song dynasty, however, Zhong and Lü—as well as the rest of the "Eight Immortals" (Zhang Guolao 張果老, Han Xiangzi 韓湘子, Li Tieguai 李鐵拐, Cao Guoqiu 曹國舅, Lan Caihe 藍采和, and He Xiangu 何仙姑)—had not yet achieved widespread popular acceptance. As a matter of fact, the so-called Eight Immortals did not step into the limelight until the Yuan period. Before then was another group of Eight Immortals, of Sichuan affiliation. Some of the members were just as august in their fame, including Zhang Daolin 張道陵, the Heavenly Master (active ca. AD 58–75) and Dong Zhongshu 董仲舒, the great thinker of the Han period (176–104 BC), but others have been cast into oblivion.[2]

1. See Xu N 1914, *juan* 6, 4b; Feng/Feng 1906, *juan* 12, [25]. Kong 1992, 733, mentions that a similar mirror of Song date was unearthed in Hunan province. That particular mirror has been published in Zhou SR 1986b, pl. 169.

2. See Li et al. 1959, vol. III, 1641, for an account of a Daoist painter, Zhang Suqing, who made a set of paintings on the subject of the earlier Eight Immortals.

83.

月宮鏡
Octafoil Mirror with Lunar Palace

Jin (1115–1234)
Weight: 954 g
Diameter: 21.3 cm
Rim height: 0.75 cm
Overall height (knob): 0.9 cm
1995.375

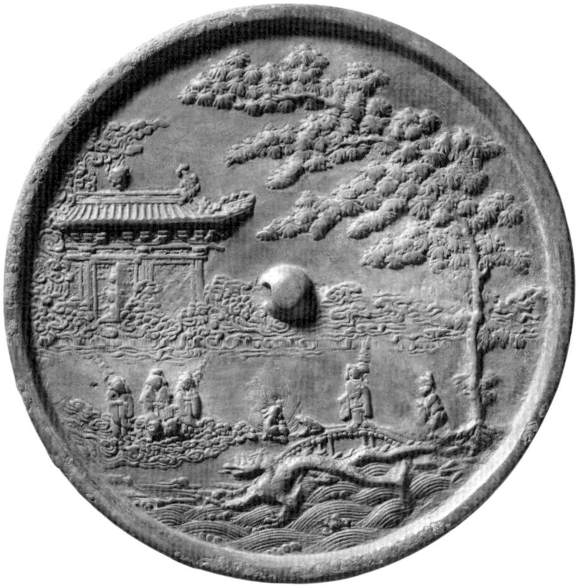

A cassia tree stands tall at the right and follows the curve of this circular mirror to tilt the tip of its foliage to the left. It nearly overshadows the palatial structure, whose door opens to reveal a figure. She may be one of the attendants to the goddess, Chang'e 嫦娥, in the Lunar Palace, who is featured at the center of the triad down below, flanked by two maidens holding parasols. Clouds envelop the building and float below to support the central knob of the moon. Underneath, on the same axis and atop the bridge, is a jade rabbit surrounded by a flaming aureole and holding a mortar and pestle to produce the elixir of immortality. A toad sits nearby, again an integral part of the myth of the moon. To the right stand two figures, one of whom holds a parasol. Below is the bridge and an undulating dragon, a curious presence on this lunar landscape.

Otherwise not a creature of the moon, the dragon may point to the dream of the Tang dynasty Emperor Ming, who upon visiting the moon was entertained with celestial music performed by resident immortals.[1] The dream itself lacks concrete details, but gave rise to varied interpretations at later times. These, in turn, assigned major roles to such immortals as Luo Gongyuan 羅公遠 or Ye Fashan 葉法善, who brought His Majesty to the lunar sphere.[2] Indeed, if the dragon is the personification of Minghuang, it is his attendant who holds the parasol above. The figure near the tree would necessarily have to be Wu Gang, a mythological fixture in the lunar scene. His frustrating task was to cut down the cassia tree, whose robust life force instantly healed the axe wounds.

Surprisingly, in a number of mirrors of this type, particularly those of the Northern Song period, the dragon does not appear.[3] When such is the case, then it is very likely that the figure near the tree (a figure which is much larger in these mirrors) must be Minghuang himself. In the Carter mirror, the reduced size and bent posture make the figure appear humble and without dignity. He is also far away from his attendant with the parasol. It may be more appropriate to interpret him as Wu Gang, not only from our point of view, but also from that of the mirrorsmith.

Such changes are not exactly unusual, but certainly require some explanation. They usually take place during the transmission of a particular theme, where continual replication over time may cause blurring, or actual mistakes, in the depiction of the respective roles played by heroes or heroines. It is quite probable that a mirrorsmith may have caused such a transformation, where the Imperial Majesty was no longer presented in such majestic tones and inadvertently became a humble Wu Gang figure. At that point, the mirrorsmith may have decided to add the dragon to the scene in order to regain the necessary link to the popular folk story. He need not have been the very one who produced this Carter mirror, but perhaps had worked earlier.

Thus, the rendition here represents a later strand within the thematic transmission. Because of this we can advocate a Jin period date, when the inclusion of the dragon began in earnest and was also being carried into Korea. This conclusion comes despite the fact that a Jin tomb too carried a mirror closer to the Northern Song lineage.[4] The changes in treatment also bring stylistic implications. In general, the Northern Song versions present the figures larger and the images closer to each other, and also closer to the viewer; the Carter version, and those like it, projects a distant vista. This latter view is far more spacious, but the figures are smaller in scale and tend to be separate from one another.

1. For the source of the Ming Huang's dream visit to the moon, see Zheng 1940, 5; Kerr 1990, 94.

2. See Li et al. 1959, vol. 1, 147, 172.

3. See for instance, Hebei 1996, pl. 290; Liang 1940–42, vol. VI, pl. 24, shows the same iconographic pattern without the dragon. For Northern Song examples, see Hefei 1991, 26–38, 70, fig. 22; and a report on the principal discoveries in Ningxia, Ningxia 1978, 58 and pl. 7-(1).

4. See Hebei 1996, pl. 290.

84.

大定通寶花紋鏡
Mirror Featuring *Dading* Coins
Jin (1115–1234), ca. 1178–1234
Weight: 139 g
Diameter: 11.9 cm
Rim height: 0.4 cm
Overall height (knob): 0.8 cm
1995.381

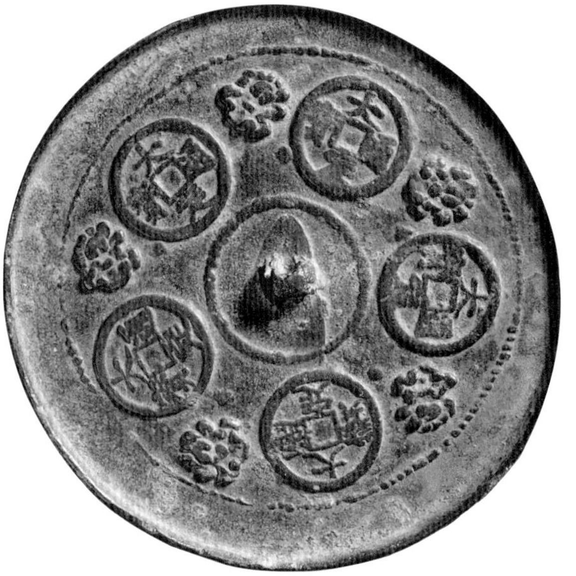

This is a simple mirror with five coins of Dading Tongbao 大定通寶 arranged in a circle around the knob, the core of this design. A centripetal force is exerted by the coins' placement, with the four Chinese characters read toward the center.

Indubitably, this is a Jin product, as the coins themselves were made during the Dading reign (1161–90) of Emperor Shizong—to be precise, in the eighteenth year of his reign (1178). Thus, mirrors of this type must postdate that year.[1]

1. Mirrors featuring coins of the Dading reign vary from those with a single coin attached to an otherwise unrelated design to those which utilize coins as the principal motif (e.g., this Carter mirror). For the former, see Hebei 1996, pl. 228; for the latter, ibid., pls. 256, 377. For a Jilin find, see Kong 1992, 872.

85.

童子騎馬鏡
Mirror with Wheel Spokes and Riding Boys
Jin (1115–1234), ca. 1196–1234
Weight: 311 g
Diameter: 13.3 cm
Rim height: 0.7 cm
Overall height (knob): 0.8 cm
1995.382

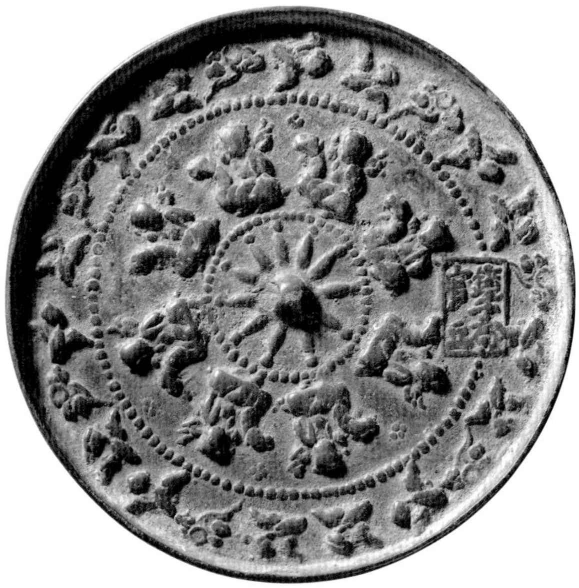

This novel design is without parallel in South China. In this circular mirror the central knob becomes the axle of a multispoked wheel, around which children on horses move counterclockwise. The outer border is filled with a flight of wild geese interspersed with cloud scrolls in a clockwise fashion; this juxtaposition creates a complex dynamic interaction. The cartouche contains the inscription 官立鏡子局 (Official Bureau of Mirrors).[1] It has been said that this bureau was established during the Cheng'an reign of the Emperor Zhangzong (1196–1200).[2] Thus we may date the mirror after 1196 but before 1234.

Amid many claims for the Jin mirrorsmith's innovations, representation of children has been one of the most prominent. Those who made this observation must have been unaware of a parallel interest in the south—namely, the Jin's nemesis, the Southern Song, where Su Hanchen made the painting of children a respectable genre. His *Children Play at the Autumn Courtyard* depicts well-behaved, refined children enjoying an innocuous game in a private garden.[3] Of course, his role as court painter incited admiration and duplication, and also made a significant impact on the Song ceramic industry.[4]

This is not to deny that the northern, Jin tradition introduced something new into the art of mirrormaking, only that the claim for innovation should be modified to suit the actual circumstance. Consequently, the only claim that the Jin mirrorsmith could make would be combining the representation of children with that of the horse, such as in this mirror. When dealing with Jin mirror traditions, one also must be mindful that the clientele included both the Jurchens and the Chinese, the ruling class and the ruled. The large populations of the latter obviously contributed to the survival of these mirrors with Sinic subjects, including those with the theme of active children depicted here, the theme of the "twin carp" (see no. 86), and even narrative subjects such as those in mirror nos. 80 and 83.

1. See Hebei 1996, pl. 305, which features the same typology and official seal and which was excavated from Wanxian, Hebei province.

2. See Yang 1995b, 113, under entry for pl. 10.

3. National Palace 1989, 67.

4. See Zhejiang 1988, 55–60, especially p. 56, figs. 3, 4-(3).

86.

雙魚鏡

Mirror with Twin Carp

Jin (1115–1234), ca. 1200–25

Weight: 813 g

Diameter: 17.9 cm

Rim height: 0.7 cm

Overall height (knob): 0.9 cm

1995.379

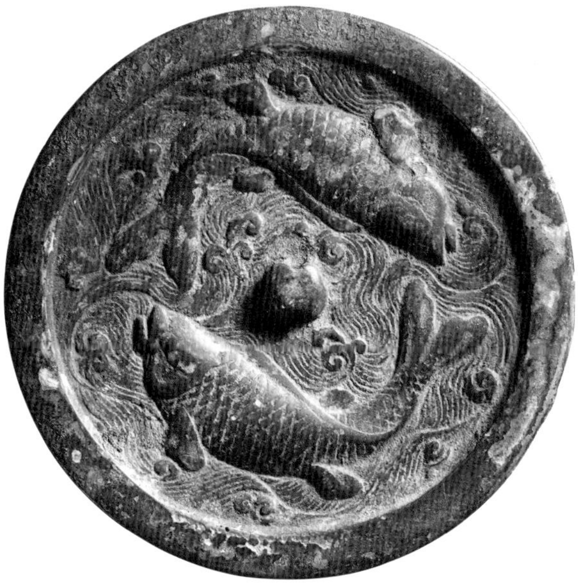

This mirror type, with two fish in inverse postures, appears to be prevalent in fair numbers in northern provinces such as Jilin, Hebei, Shaanxi, and Henan. The province of Hebei has yielded the most examples—either through excavation or through collecting in the marketplace. Regardless of where they were discovered, the provinces where these objects originated were under the control of the Jin dynasty during much of the twelfth and early thirteenth centuries. Thus, a Jin provenance is usually attached to this type of mirror.[1]

The reign of Cheng'an (1196–1200), of Emperor Zhang of the Jin dynasty, produced dated examples. The mirror in plate 244 in *Li dai tong jing wen shi* 歷代銅鏡紋飾, a volume dedicated to Hebei finds, has an inscription that gives the date of the second year of that reign, 1197, and shows that it was a production piece by the Bureau of Mirrors. Another, dated in the fourth year of the reign (1199), identifies the maker, Yang Lin, as a craftsman in the official workshop located in Shaanxi Eastern Circuit and under the control of the chief transport commissioner.[2] This second mirror carries a design of concentric circles, with the twin fish in the inner circle, and a ring of inscription in the outer.

While twin fish is a typical Jin motif, the fact remains that the craftsmen of these mirrors took their cue from Song mirrors of similar design. The leaping carp, as rebus for political success, was long known in the Chinese world before the idea spread elsewhere. A mirror of Song origin displays, very much like the 1197 specimen, the basic design pattern where the fish are smaller in proportion to the surface they occupy, portrayed in simple profiles with no twist in their bodily axis. The 1199 mirror expands the bodies of the fish to conform to the circular contour. The twist of their tails is not dissimilar to the representation of fish in the Sinic sphere, but lacks the lithesome movement of the Song examples. One might be tempted to conclude that these mirrors of the Cheng'an were indeed the initial attempts by Jin craftsmen, whether Chinese or Jurchen, to work with a subject that for one reason or another was being turned into a major thematic element in the north, so much so that some are tempted to characterize such mirrors as a "typical" product of the Jin.

In comparison, the Carter mirror represents a full embodiment of the swimming carp theme. These carp are very much alive—strong and powerful creatures swimming boldly in swirling currents and foaming waves. The knob is small and the border is plain, placing the focus on the dynamic fish. One suspects that decades may have passed from the time of the 1190s' specimens before the mirrorsmith acquired sufficient skill and imagination to create this piece. Accordingly, a dating of 1200–25 may be suitable.

1. A twin-fish mirror was also said to be from a Liao tomb in Yueyang, Hunan; see Zhou SR 1986b, pl. 191.

2. First published in Liu T 1935, *juan* 17, 89a, it also appears in Kong 1992, 816. Another mirror of the same type can be seen in Taipei 1986, pl. 150.

Phase V

Yuan, Ming, and Qing Dynasties 1279–1911

87.

雙龍荷花鏡
Mirror with Twin Dragons and Lotus Blossoms
Yuan (1279–1368), dated 1338
Weight: 1,676 g
Diameter: 23.4 cm
Rim height: 0.8 cm
Overall height (knob): 1.1 cm
1995.388

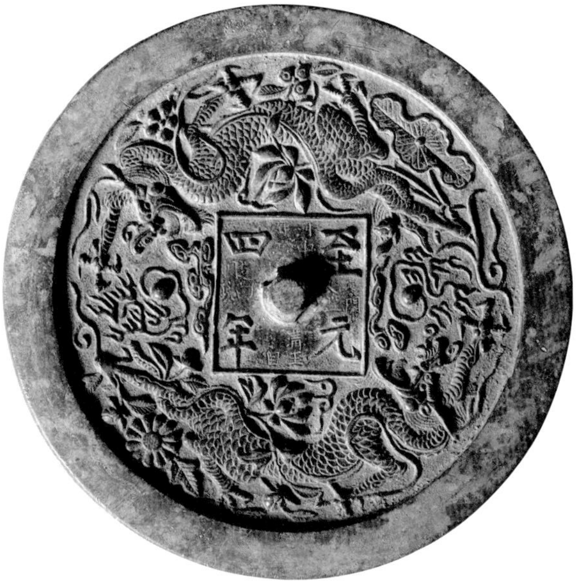

This mirror features a small knob, flattened at the top, and surrounded by a square enclosing the four characters of 至元四年 (Fourth Year of Zhiyuan, 1338) in relief. Among these bold characters are a series of sixteen engraved characters in regular script, barely legible, that establish the identity of the workshop concerned: 河中府馬溝玉家自家門人監造白銅鏡 (White bronze mirror cast under the supervision of retainers from the house of Ma Gouyu in the prefecture of Hezhong).

Hezhong refers to the prefecture at the point where the Yellow and Fen rivers meet, in the province of Shanxi. This was where Ma Gouyu had his workshop. The master probably was in his later years, so he entrusted the task to unnamed disciples who supervised the casting.

The twin dragons are the most notable images, circulating in concert with the inner border of the plain rim, ready to swallow the flaming jewels. Lotus, chrysanthemum, peach, and loquat blossoms also make their appearance in whatever space is left by the dragons.

A number of mirrors share this same pattern of design. An archaeological site in Handan, Hebei, yields such an example,[1] and another was found in Liaocheng, Shandong.[2] Were all these the products from the Ma family workshop? In answer to this question, the mirror unearthed from Zhangxian, Gansu province, bears critical evidence. It was reportedly made in a Chang'an workshop.[3] If so, it would appear that workshops at that time copied from one another, or perhaps derived their designs from an official workshop. One can also wonder if it were possible at that time to have had several franchises permitted to produce the same design.

1. See Hebei 1996, pl. 388.

2. See Liaocheng 1993, 90–93, fig. 10. The diameter of this mirror is only 22.5 cm, and the rim is 0.4 cm. It is a trifle smaller and thinner than the Carter version, but heavier.

3. See Gansu/Zhangxian 1982, 13–21, especially p. 17, fig. 6. This mirror measures 22.5 cm in diameter, with a rim height of 0.5 cm.

88.

蓮花鏡
Mirror with Lotus and Mantra
Yuan (1279–1368)
Weight: 787 g
Diameter: 22.3 cm
Rim height: 0.7 cm
Overall height (rim and convex reflective surface): 1.5 cm
1999.221

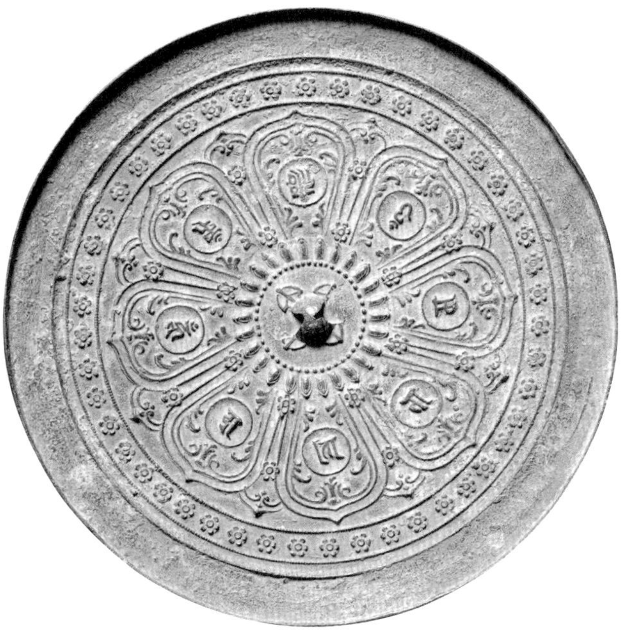

Cast in low but sharp relief, this is a Buddhist mirror with a clear identity and a precisely arranged sequence of design motifs.[1] A domed knob is at the center, surrounded by four leaves and a circle of beads, which in turn is encircled by twenty-six miniature lotus petals. Beyond, eight broad petals are fully visible, with eight more in between. Each broad petal has dual contours—within them is a circle, designed to enclose a Sanskrit letter—whereas the overlapped petals are decorated with two prunus blossoms. A ring of plum blossoms, plus two plain concentric rings, complete the design. The rim's wall is thin and high, and the mirror itself, seen from its reflective surface, is notably convex.

Predictably, the mantra contains the familiar *Om mani padme hum.* The Sanskrit here is in a bastardized form, however. We speculate that this was largely due to the Chinese mirrorsmith's unfamiliarity with that language. The full translation of the text is: "In the name of *Om mani padme hum*."[2]

1. A similar Buddhist mirror in the collection of the History Museum, Beijing, does not have Sanskrit letters but has been dated to the Song period; see Yang 1994, 126–44, fig. 11. A better comparison is a Yuan specimen in Hebei 1996, pl. 398. The absence of archaeological support for this dating should be noted.

2. Dr. Stanislaw Czuma of the Cleveland Museum of Art and Professor John Huntington of Ohio State University graciously consented to decipher this mantra.

89.

為善最樂鏡

Mirror with Confucian Maxim

Ming (1368–1644), 15th century
Weight: 71.5 g
Diameter: 8.3 cm
Rim height: 0.45 cm
Overall height (knob): 0.6 cm
1995.395

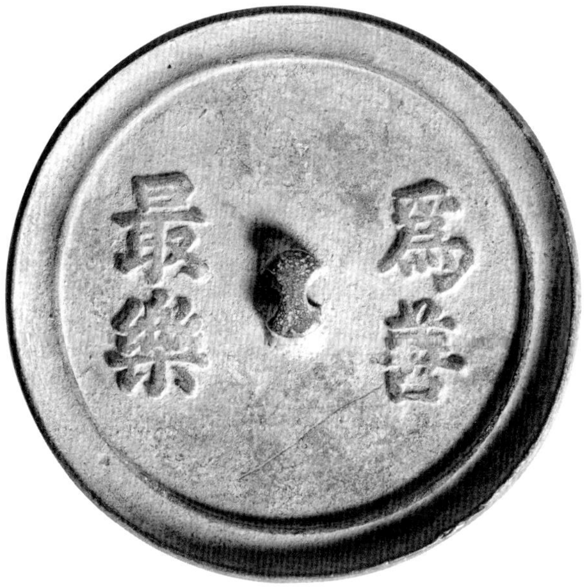

A simple mirror with green patination, its plain surface is decorated by only one circular raised ring and four characters in *kaishu* (regular script): 為善最樂 (Joy in performing good deeds). The knob has been described as a "silver ingot," which broadens at two ends and curves inward in the middle.

A published example of this mirror type can be found in *Zhong guo tong jing tu dian* (An Illustrated Compendium of Chinese Bronze Mirrors),[1] which features the well-known aphorism. More importantly, the 1991 excavation of Luo Hengxin's tomb (d. 1457, buried 1459) in Dongguan, Guangdong, yielded a mirror with the same inscription, which testifies to a fifteenth-century date for its production.[2] Luo was a court official and attained the final rank of Grand Master for Thorough Counsel and Vice Censor in Chief at Left. Another was unearthed in Pingwu, Sichuan, in a tomb datable to 1512.[3] Considering these two *termini ante quem*, and the peripheral nature of the provinces concerned, such mirrors must have been quite in vogue during the fifteenth century.

The well-executed calligraphy is in a style that could be considered a Ming approximation of Su Shi's style. Thick strokes and bold form are characteristic, with occasional defiance of otherwise tectonically structured characters.

1. Kong 1992, 929.

2. See Guangdong/Dongguan 1991, 38–50, fig. 12. This is said to be the first instance where this type of mirror appeared in Guangdong.

3. Sichuan et al. 1989, 1–42, fig. 124. The tomb was that of Wang Wenyuan and his wife.

90.

湖州素鏡
Huzhou Mirror
Ming (1368–1644), 16th century
Weight: 708 g
Diameter: 15.8 cm
Rim height: 0.9 cm
Overall height (knob): 1.1 cm
1995.370

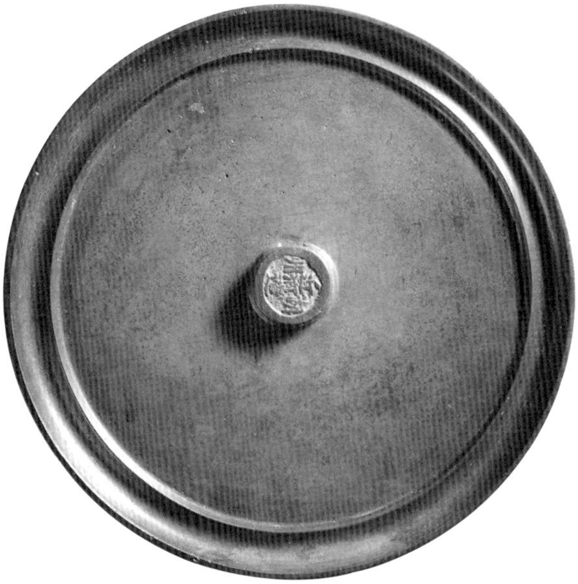

This is the only Huzhou mirror in the Carter collection. It is in a fine state of preservation, revealing a muted, grayish blue sheen. The mirror is devoid of decoration apart from its cylindrical knob, the top of which has an inscription that identifies only its site of production: 湖州近河造 (Made in Jinhe, Huzhou).

Already a center of mirror-making in the late Northern Song (960–1127), the Huzhou tradition continued uninterrupted into the Qianlong era (1736–95). It is one of the longest running of any such centers, including Shaoxing and Yangzhou in earlier centuries.

For dating purposes, comparative study with archaeological finds is essential. The closest to it in form is a mirror made by Xue Huaiquan 薛懷泉. In both examples, the decoration of the mirror back consists of only two raised rings, the outermost of which is a beveled rim. Both mirrors have a short cylindrical knob, the top facet of which is shaped like a circular seal, with an inscription, in relief, squeezed into its limited space. Unfortunately, Xue Huaiquan's mirror is poorly reproduced and thus it is difficult to discern its current condition.

Xue Huaiquan's mirror was unearthed from a Ming tomb in Xuanhua, Hebei province.[1] Another find, in Suzhou, took place in the tomb of the famous scholar and official, Wang Xijue 王錫爵 (1534–1610), and his wife (d. 1598).[2] This second mirror, dated 1591, allows us to arrive for the first time at a rough chronological framework for the master craftsman himself, and by extension, that of the Carter mirror.

What remains, of course, is the question of which of the two mirrors was to provide the original inspiration for this simple mirror design. Xue, from what we know today, was not a Huzhou native.[3] He based himself in Kaihua, and may have been derivative of the Huzhou tradition. If this is the case, instead of Xue being the original source for this type of mirror, it was the Carter version, or mirrors like it, that had inspired him. We are, in short, witnessing a phenomenon that saw mirrormakers elsewhere gravitating toward the famed center and reproducing what it had to offer.

1. See Hebei 1996, pl. 422.
2. See Suzhou 1975, 51–56.
3. See Wang SL 1987, 18–23, especially p. 23. For another survey of the Huzhou mirrors and several other local traditions, see Wu S 1993, 11–15.

91.

四乳四虺鏡
Mirror with Four Nipples, Four Quasi-Dragons, and Birds
Ming (1368–1644)
Weight: 355 g
Diameter: 10.5 cm
Rim height: 0.75 cm
Overall height (knob): 1.1 cm
1995.393

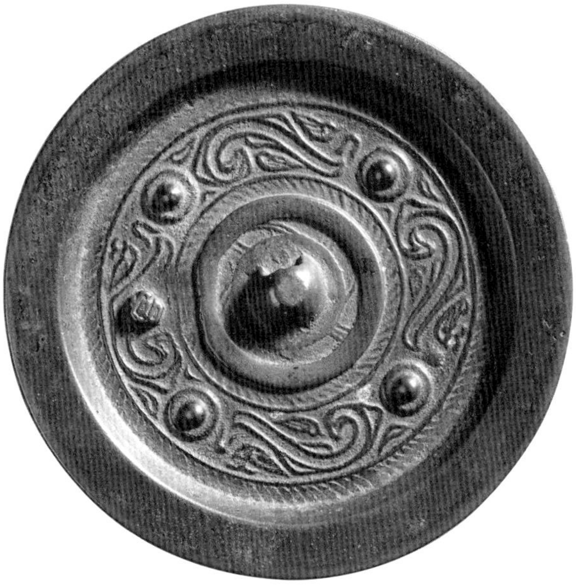

This mirror is without doubt a recast of a Han mirror (see no. 20). The multilayers of concentric circular bands and the broad unadorned rim are typical of early mirrors, as are the four nipples and the undulating organisms in between with smaller birds above and below. The only addition is the character *gong* 宮, cast in relief and at an angle that cannot be said to coordinate with any visible axis. Does this refer to mirrors used in the Imperial Palace? The quality of the mirror does not seem to warrant a court connection. Could it point to the surname of a craftsman who was responsible for having reproduced at least two mirrors after the Han antecedents?[1] That is quite possible. Examples of Master Gong's mirrors have survived; mainly of fully developed TLV variety, they have a small character of *gong* in a circle either between a set of the T and L bars, or along the band of inscription. In one, the full name of the craftsman is unveiled: Gong Xiaoshan 宮小山. In another, the location of his shop is said to be Yizhen 儀真. The main question remains, when were they made?

One scholar, Yang Guirong, who has catalogued the mirror collection at the Museum of National Chinese History, Beijing, contends that these mirrors were Song imitations of early mirrors.[2] His reasoning, it must be said, lacks archaeological backing. He may also have neglected to research the location of the Gong workshop, Yizhen, within the Yangzhou prefecture. Had he done so, he would have discovered that it was so named only from 1369 onward, i.e., the second year of Hongwu, the reign of the first emperor of the Ming dynasty.[3] That discovery compels one to assign these Gong mirrors to the Ming period. Because of this, the Carter mirror shall be given a Ming date.[4]

1. See Yang 1995a, 135–43, pls. 84, 86; for explanatory note, see p. 137. For two additional Gong mirrors, see Liu T 1935, *juan* 17, 44b, 45a.

2. See Yang 1994, 126.

3. See Zhang T et al. 1967, *juan* 40, 28566–7, under Yangzhou Prefecture.

4. Another mirror of the same type has been dated to the Ming period; see Lüshun 1997, pl. 209. This means that the same type of mirror was made in more than just one workshop.

92.

雙喜五福帶柄鏡

Mirror with Handle, Decorated with "Double Happiness" and Five Bats

Qing (1644–1911), ca. 1800

Weight: 310.5 g

Dimensions: L 26.7, diam. 16.5 cm

Rim height: 0.6 cm

Overall height (rim): 0.6 cm

1995.401

(Plate 8)

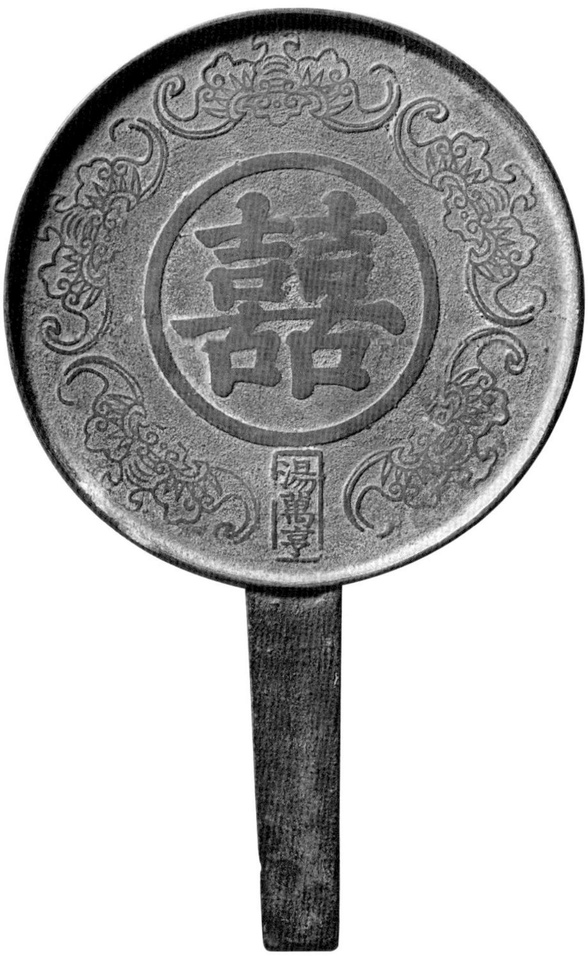

This mirror's golden reflective surface indicates that the material is brass, whereas its back is intentionally darker and more somber. The back also bears a large character of *xi* 囍 ("double happiness," as in a marriage) as its principal decoration. Below that, a smaller inscription in a rectangular frame establishes the mirrorsmith's identity. Tang Wanheng 湯萬亨 must have enjoyed a lucrative trade and sufficient renown to warrant the inclusion of his name on his products. It has been suggested that these mirrors were not so much made for reflective purposes as for ritualistic use during the marriage ceremony, hence the "double happiness" as its main decoration. The five bats also signal the onset of good fortune.

Other than this Carter specimen, several surviving examples have been published. *Xiao jiao jing ge jin wen tuo ben* (Rubbings from Bronze Artifacts and Inscriptions in the Xiaojiaojing Pavilion)[1] reproduces three examples—two of which contain the names of the makers. One has the inscription "I, He Sheng, made this precious mirror"; Gao Yushan made the other piece. *Zhong guo tong jing tu dian* (An Illustrated Compendium of Chinese Bronze Mirrors) brings to our attention an excavated example from Changsha, Hunan province.[2] However, the most important is the mirror from Xushui Xian, Hebei province. It contains the inscription "Made by Xue Jinhou of Huzhou." Master Xue, also known under the name of Xue Huigong, was active in Huzhou during the eighteenth century. He was perhaps the most celebrated mirrormaker of his time—before the abundance of glass mirrors overturned his fortune.[3] It is possible that he initiated the design, which then was copied by lesser artists or workshops, including Tang Wanheng's. By the same token, the root of this mirror type lies also in the Huzhou heritage despite the broad geographic spread as seen in the above examples.

1. Liu T 1935, vol. 17, 108b, 109a–b.

2. Kong 1992, 943.

3. Wang SL 1987, 18–23, especially pp. 22–23.

Appendix

Technical Issues

Making the Mirrors

THE BASIC FORM OF THE MIRROR as a cast metal object with a highly polished reflective surface and a back with relief decoration has remained unchanged throughout most of China's history. The earliest known Chinese mirrors date to approximately 2000 BC. Over the next 3,500 years there was variation in decoration and composition, but not in the mirror's essential form. Highly polished bronze gave way to silvered glass mirrors in the Qing dynasty during the 18th century.[1] The 146 mirrors in the Carter collection—which spans the time period from the Warring States (480–221 BC) to the Qing dynasty (1644–1911)—provide a rare opportunity to understand the changes that took place in mirror making. The most obvious are in decorative motifs. Other, more subtle changes occurred in the method of manufacture, of casting and finishing, and are more difficult to discern. The earliest mirrors were created using clay molds, a technique that gave way to the lost-wax method. Finishing work on the clay mold mirrors was largely restricted to the top of flat design areas or on knobs, nipples, and linear channels. On the lost-wax cast mirrors, finishing work became much more extensive, with nearly all surfaces smoothed. Changes occurred in the composition of the bronze alloys over time, often depending on the availability of tin and on changing economic conditions.

Each mirror in the collection was examined using a variety of techniques. Visual inspection was followed by examination under magnification up to 50x using a Wild binocular microscope. Evidence such as mold cracks and other damage, modeling, the type and degree of finishing work, and sharpness of the cast design was used to identify manufacturing methods and mirror condition. X-radiographs also helped determine condition, as many of the mirrors have internal cracks visible only in the x-radiographs.[2] Six x-radiographs were very dramatic, showing extensive restoration. The mirrors were also examined with energy-dispersive x-ray fluorescence spectrometry to obtain information about metal composition.

THE CASTING PROCESS

The mirrors in the Carter collection were cast either in preformed clay molds or in molds created by the lost-wax method. At its simplest, the use of clay molds involved creating a void in the clay in the shape of the mirror and then pouring in the molten bronze. The lost-wax method at its simplest involved creating a wax model, encasing it with clay, and then firing it. The wax melted and created a void in the ceramic in which to pour the molten bronze.

Bronze casting has a long history in China, dating back to at least 3000 BC. Historically, the Chinese preferred casting rather than hammering for shaping metal. Initially, molten bronze was poured into complex ceramic molds to create decorated bronze vessels. The making of these complex ceramic molds suggests that the bronze industry must have grown out of or in conjunction with the ceramic industry in China. The earliest mirrors were made using clay mold technology. The lost-wax method of casting came to China later, not arriving until the Han dynasty (206 BC–AD 220). In comparison, the lost-wax method was used in Egypt and the Middle East by 2500 BC. The use of clay molds for making mirrors continued after the introduction of the lost-wax method for casting other forms such as vessels and weapons. Lost-wax technology for making mirrors was not used until the Sui and Tang dynasties (618–907).

The technical proficiency of the foundrymen from an early date is evident. From the way in which the molds were made and the alloys used, it is clear that they understood the problems involved in casting metals. In designing molds to cast complex objects they learned to create vents to avoid cracks in the objects caused by shrinkage and to minimize porosity. They learned to vary the composition of their bronzes depending upon the object's intended use. For example, more tin was used to create harder edges for weapons, while other alloys used more copper for ductility.

The foundrymen carefully chose the composition of the bronze alloy they used for casting the mirrors. The mirrors are generally made of copper, tin, and lead. Adding tin to copper lowers the melting point of the alloy below the melting point of copper.

The lead helps the alloy to pour more easily during the casting process and facilitates grinding and polishing. They also exploited the color of a high-tin bronze to create a silver-colored reflective surface. Over the years Chinese mirrors have been analyzed by a variety of techniques. Modern analyses show that the composition of the mirrors varied with time and place, probably due to economic conditions and the availability of tin. In general, the tin content in the earliest mirrors before the Warring States period was often low and variable. Mirrors from the Warring States period through the Tang dynasty had a fairly consistent alloy of 66 to 78 percent copper, 18 to 26 percent tin, and 1 to 9 percent lead. After the Song dynasty the tin content declined (in some cases to as low as 2 percent) and the lead content increased, often to 20 to 25 percent. A well-known set of "Six Formulas" from *Zhou Li* provides recipes for various bronze forms.[3] As W. T. Chase notes, the formula for mirrors is given at 50 percent copper and 50 percent tin, which does not agree with the analytical data. Chase feels the discrepancy between the formulas and the actual analyses comes from unstated practices. Lead was always left out of formulas but was clearly used.

The skill of the Chinese foundrymen over the centuries is revealed in the Carter collection. The x-radiographs rarely show trapped gases in the cast metal; porosity was rarely found on the surface. The decoration on each mirror was cast without obvious flaws, with no major repairs made at the foundry visible. Problems occasionally did occur. Sometimes the mold would crack before the metal was poured. The metal would flow into the crack, filling it with a ridge of metal known as a casting fin. Occasionally part of the mold would break away before the metal was poured. In these cases, the metal would fill the void left by the lost section. Sometimes the foundryman did not have the metal or mold warm enough, causing the molten bronze to solidify before properly filling all the sections. Such premature solidification results in "chill flow."

CLAY MOLD TECHNOLOGY

To make a mirror using a clay mold, a void in the clay is created in the shape of the mirror and the molten metal is poured into it. Several of these clay molds have been discovered in China.[4] The molds were made in two sections: one side contained the decorative back and the other side was flat. Channels for pouring in the molten metal and for escaping gases were made at one end of the mold. The dried molds were heated before the metal was poured. The molds had to be dry to prevent the generation of steam, which would have caused gas in the casting.

While the clay mold casting method is relatively simple, details on how the molds were created are not entirely understood. Since casting is a reproductive process, a wide range of variation in the mold-making procedure is possible. There are two fundamental ways to create a clay mold. The first method involves creating a model of the mirror in some easily formed material such as clay or wood. A clay mold for casting the bronze would then be fabricated by packing the clay around the model and then removing it. The second method requires creating the mold in the negative. In this method, carving or scoring the clay with sharp tools or ceramic stamps creates the decoration on the mirror. Obviously these techniques could be combined to create the final decoration. It is generally accepted that clay molds were used to make most bronze vessels. However, Meyers and Holmes demonstrate that the vessels were made using a model, creating the design and transferring it to the mold.[5] In another example they show that a stamp was used to press the decoration directly into the clay mold. The archaeological record and literature do not shed much light on which method was used to create the clay molds for mirrors. In examining the Carter collection, no evidence arose that negative mirror molds were made or that there was much modification of the design within the mold. The easiest and most efficient way to create a clay mold would be to make a model. It is also much easier to visualize the object as a positive rather than a negative image. A significant number of mirrors made in clay molds have small clumps of metal in the designs. This type of damage suggests the use of a model that was pulled away from the mold.

During the Han dynasty, stone molds were occasionally used.[6] The design for the decoration was carved into the stone and—like the clay mold—consisted of two sections. The advantage of using stone was that the foundry could reuse the molds multiple times, whereas each clay mold produced only one mirror.

LOST-WAX TECHNOLOGY

In the lost-wax method, the mirror was first fully modeled in wax to create the exact form of the final bronze mirror. The wax could be built up, carved, heated, or dipped to shape. A wax funnel (sprue) and small tubes (vents) would be attached to the model—a means to pour molten metal into the mold while allowing a way for air to escape. The wax model with its system of sprues and vents was then invested—that is, encased in clay or clay-like material—and fired. The firing did several things. First, it hardened the clay mold. Second, it melted or burned the wax out of the clay, leaving a void. Finally, it heated the mold, which helped to keep the molten metal from solidifying prematurely. After the metal solidified, the mold was destroyed, leaving the bronze mirror.

Study of this collection suggests that wax models were created in several ways. Two Tang dynasty mirrors (nos. 66a, 66b) provide an interesting insight into wax model-making practices. The designs on their backs are identical in size and placement; the only real difference is the size and shape of the border. Both mirrors were made by pouring molten wax into the same mold, then customized by adding a different size and shape border. Many forms on various mirror backs seem to have been cast in similar molds. Alternatively, they may have been made with sprig molds. Sprig molding involves pressing or casting the wax into a shallow relief mold, then removing the shape and attaching it to a sheet of wax in the shape of the mirror. In this way, the mirrors could be constructed with great variety (nos. 62–64, 67, 71). The process of making lost-wax models is well suited to such a working methodology.

Cruciform Mirror with Two Lions (no. 74) provides another insight into the model-making process. As in the two Tang mirrors discussed above, each border appears to have been made separately and attached to the mirror. In this particular case, the animals were probably sprig molded and attached to a wax background. The rings in the wax background appear to have been created by poking a tube into the back of a thin sheet of wax. *Mirror with a Coiling Dragon* (no. 72) shows that molten wax was dripped onto the dragon shape to create scales. The *Huzhou Mirror* (no. 90) has an iron chaplet on its reflective side that is virtually unique. A chaplet is a piece of metal that was used to keep the wax and molds in alignment. Its presence in this relatively undecorated object suggests that the mirror may have been mass produced. A number of identical wax mirror models may have been stacked together on the sprue. The chaplet provided extra strength to keep all the wax models and clay in place.

Mirror designs were influenced by available technology. Decoration on earlier mirror backs was restrained by clay mold technology. The decorations on the back of the earliest mirrors (Warring States and Qin) tend to be done in low relief. This may be a result of the problems that occurred while pulling the mold away from the model. Small parts of the mold would break away when the model was removed. Such damage to the mold is evident in a number of mirrors, causing lumpy areas. Later, particularly during the Eastern Han, the relief decoration tends to have greater depth and at times is nearly three-dimensional. Perhaps the craftsmen became more comfortable and experienced in working with the clay and separating the model. This desire to create three-dimensional decorations may have been the impetus for craftsmen to finally adopt the lost-wax technology that had been in use in China since the Han dynasty for the creation of vessels, weapons, and other utilitarian objects. The lost-wax process gave greater creative freedom by allowing more fluidity, depth, and variation.

DISTINGUISHING ONE TECHNOLOGY FROM THE OTHER

Determining whether a mirror was made by the lost-wax process or with clay molds is not an exact science. The mirror's surfaces must be examined closely for evidence of mold cracks or other damage, modeling, type and degree of finishing work, and sharpness of the cast design. Some features typically seen in objects made by one casting method can and do show up in objects made by the other method. In general, mirrors cast in ceramic molds tend to have crisp designs and features. On the other hand, designs on mirrors made by the lost-wax method tend to be softer and more three-dimensional. Occasionally mirrors would be recast at a later date by the lost-wax method, producing designs that lack the original crispness.

As discussed in more detail below, finishing work on mirrors cast in ceramic molds tends to be limited to polishing of the raised flat, level design areas, chiseling of concave linear design elements, and rotary abrasion of depressions, nipples, and center knobs. Lost-wax cast mirrors tend to be extensively finished; chiseling and polishing of the background often continues into the relief decoration, resulting in some softening of the design.

Mirrors cast by ceramic mold technology often will have some imperfection caused by mold damage. Two types of damage are common. Part of the clay mold can break away before the metal is poured, resulting in an irregular clump in the cast design. Cracks in the mold can produce a metal ridge called a casting fin. These two features can also occur in the lost-wax method, so are not definitive. A clump of metal from a loss of mold material was occasionally observed in some of the Carter collection's lost-wax mirrors from the Tang dynasty and later. Evidence of mold cracks is extremely rare, perhaps because the extensive finishing work would have removed it.

FINISHING WORK

No matter which casting method was used, every mirror required finishing work. Even near-perfect castings have a finely pebbled metal surface, a result of the mold texture and the solidifying process. The mirror's reflective surface would be smoothed with scrapers and chisels to remove any imperfections and then carefully polished with abrasives to produce a surface that reflected light easily. The high tin content of many bronze mirrors created a silver-colored surface.

Finishing the three-dimensional decoration on the back was more difficult. The Carter collection provides the opportunity to observe changes in finishing practices over time. Visual and microscopic examinations reveal the extent of finishing work. The excellent state of preservation allows detection of chiseling and other tooling. Casting imperfections such as cracks and clumps of metal as well as the as-cast pebbled surface observed in the relief decoration show a lack of (or at least minimal) finishing work.

The cast designs of the clay mold mirrors were usually sharp and crisp. Their complexity made finishing work difficult. The raised flat areas were scraped smooth and polished almost as finely as the reflective surface. Concave linear elements in the TLV designs were often scraped out with a chisel that was the width and shape of the concave channel. Parallel striations from this process generally remain in the metal. Circular abrasion marks on the mirrors with nipples indicate that a drill—most likely some type of bow drill—was used for finishing work. A wood or bamboo drill bit with a negative nipple shape at one end was rotated in the drill with an abrasive to polish the nipples and center knobs. Circular abrasion marks in dimples suggest use of similar finishing methods. Rarely is any finishing work apparent in recessed areas of the designs.

Mirrors created by the lost-wax method also required finishing work. During the Sui dynasty abrasion was the principal finishing technique, used mainly on the flat spaces between raised areas of the decoration, leaving the pebbled casting surface in the adjacent background area. Sometime during the Tang dynasty craftsmen used flat chisels to scrape and smooth the vast expanses of flat background; tool marks often are visible near the relief decoration or on the borders. Scraping usually yielded a slightly rippled metal surface even after polishing. The relief decoration was often polished, resulting in some softening of the details. Close examination of these mirrors reveals a rough pebbled surface from the casting process in background areas immediately adjacent to the relief decoration. These areas were too tight to smooth effectively—either by chiseling or by polishing—without excessively damaging the relief decoration.

COLOR OF THE MIRRORS

The original appearance of the mirrors is not fully understood. Many of them have well-preserved surfaces that are silver to silver-gray in color. Based on the high proportion of silver-colored mirrors in the Carter collection, it seems likely that the craftsmen often intentionally exploited the natural color of the high-tin alloys. There are many references to the use of *xuanxi* to polish mirrors to a bright silver color, which also suggests that most mirrors were meant to be silver-colored. *Xuanxi*—a mixture of tin, mercury, alum, and deerhorn ash[7]—was used to polish the mirror to a white-colored surface by leaving a thin coat of tin. Over time the coating would oxidize, necessitating repolishing with *xuanxi*. It is much debated whether the mercury stayed on the mirror surface or vaporized with time or heat. It should be noted that mercury not associated with the red pigment cinnabar (mercury sulfide) was detected on a few mirrors.

Some of the mirrors (e.g., nos. 66a, 72) have a dark, lustrous patina. This coloration—especially the black patina—has been widely debated and is not yet fully understood. The question is whether the color is intentional or a result of burial. The ancient literature has shed some light on the subject, but questions remain. Sources from the Song period identify four different colors for mirrors: black lacquer antique (*heiqigu*), green lacquer antique (*lüqigu*), quick-silver white (*shuiyinqin*), and lead gray (*qianbei*).[8] It is not clear if these classifications refer to excavated mirrors or to mirrors being produced at that time. The issue is further complicated by the use of *xuanxi*, which translates as "black tin." It is clear from the literature and experiments that this polishing compound is not the direct cause of the black mirrors. It was used to produce a bright silver color.

A number of scientific studies examining microstructure and chemical composition have been undertaken to better understand bronze mirrors and their coloration. Most studies agree that the surfaces of excavated mirrors have been altered by corrosion and tend to be rich in silica. Analysis of the data, of course, can lead to different interpretations. Zhu Shoukang and He Tangkun contend that the various patinations are a result of corrosion caused by burial.[9] They believe that the mirrors were periodically polished with *xuanxi*, the composition of which could vary. Different colorations would result during burial based on the *xuanxi* composition and the burial environment. They note that the soil in North China is acidic while that in the south is basic. Black, green, and gray-colored mirrors are excavated in the south, while bright metal mirrors are found in the north. In contrast, Chase et al. believe that the mirrors were intentionally treated to produce the colors.[10] They note that the microstructure and composition of the surface layers produced by the intentional treatment are very similar to a naturally corroded surface.

It is not our intent to settle these issues here. Instead, observations about color will be made regarding several mirrors in the collection. Some of the mirrors clearly show intentional coloring of the metal. The maker of mirror no. 25 used contrasting metals to produce a multicolored back. The center area is gilded and the silver-colored zigzag border is surrounded by intentionally applied black triangles of unknown composition. In mirror no. 26, the four animals in each corner of the design are clearly silver-colored in relation to the background. Microscopic examination shows that some material was brushed onto these areas to create the enhanced silver color. Finally, mirror no. 9 has a gold-colored background in the area of the whorl pattern which contrasts with the silver color of the adjacent bands. Examination of the gold-colored surface with energy-dispersive x-ray fluorescence spectroscopy reveals no presence of gold. In fact, there is no discernible difference in composition.[11] While it is possible that the gold-colored areas are the result of conservation treatment, there is no evidence for this. Instead the gold whorl areas are nearly free of corrosion, whereas adjoining polished bands are heavily corroded. This evidence suggests that the whorl areas were treated in some way. There is much green corrosion on the adjoining polished bands but no corrosion on the colored field. Two other mirrors (nos. 13, 28) were clearly altered by the foundryman to create contrasting colors as a part of the decoration. On these mirrors the raised parts of the design are silver, while the recessed areas are gray. The contrast corresponds exactly with the decoration and appears to be intentional, or at least the happy result of contrast between polished and unpolished areas.

THE EFFECTS OF BURIAL

Centuries of burial can greatly change a bronze mirror's original color. Air, water, and chemicals in the soil act upon the bronze, transforming the metal to oxides and carbonates. Often one alloy component will be preferentially removed from the metal. For example, a variety of copper corrosion products occurs on bronzes, such as cuprite (red), malachite (green), and azurite (blue). Tin oxides, often black or gray in color, will also form during burial. This buildup of corrosion products creates a mottled, uneven surface. Pustules—localized areas of corrosion that erupt from the surface—often form, especially on the reflective side. The corrosion process may also produce a fine network of cracks in the reflective surface.

While these changes in appearance are common, it is not surprising to find that many mirrors are very well preserved with much of their original silver-colored metal visible. This is especially true of earlier mirrors, which have a high tin content that protects against corrosion by forming a passive tin oxide layer.[12] In addition to providing a silver color and corrosion resistance, the high tin content creates an alloy that is very hard and brittle. Many of the mirrors in the collection are cracked or broken. Quite often the corrosion contributes to cracking. In a small but significant number of mirrors, the cracks can only be detected in the x-radiographs (e.g., mirror nos. 2, 11).

Restoration and cleaning of the mirrors after excavation also dramatically changes their appearance. Harsh chemicals and electrochemical stripping methods often remove the green, blue, and (to a lesser extent) red corrosion products. These treatments may be successful in revealing some of the cast details, but the mirror's surface is often marred by redeposited copper or pitting. In other instances the corrosion products are scraped or ground off, scratching the reflective surface or the decorative back. Broken or severely cracked mirrors are often restored with lead-based solder or glue. Missing areas can be built up with plastic putty or a lead-based alloy. Several of the mirrors in the Carter collection (nos. 5, 32, 33, as well as CMA 1995.269, 1995.271, and 1995.272) have been entirely covered with paint to hide the damage and the extent of restoration.

In the Appendix, the color of the corrosion observed on each mirror is given, rather than the crystal name, since it is beyond the scope of this project to sample all the different types of corrosion and run x-ray diffraction patterns for each. It is difficult to determine the type of corrosion visually, especially tin oxides.

Many of the earlier mirrors were buried inside or in direct contact with textiles. Some of the textile fibers, which are generally very small and in poor condition, are occasionally preserved in the corrosion layers. In many cases the corrosion layers preserve the weave pattern, known as a textile pseudomorph. When the weave pattern is visible on a mirror, this has been noted in the Appendix entry.

NOTES ON MEASUREMENTS AND WEIGHTS

A. G. Bulling and Isabella Drew showed that the ratio of weight and diameter could be used as an aid in dating Chinese mirrors.[13] This becomes particularly important in later periods, when craftsmen were copying or recasting mirrors from earlier periods. During the Sui and Tang dynasties, Han mirrors were copied or recast. The Sung "picture mirrors" of the Eastern Han and the grapevine and lion mirrors of the Tang period were also copied or recast. To aid identification and dating, each mirror in the Carter collection was measured to the nearest millimeter in all three dimensions. The largest diameter (or width and length) was recorded for each object. The greatest depth (often the center knob) also was recorded, as was the thickness of the rim. Mirror weights of less than 2,150 grams were measured to the nearest 0.5 gram. Mirrors heavier than 2.1 kilograms were weighed to the nearest 0.02 kilogram.

Identifying later copies also requires careful examination of manufacturing method and alloy. The designs on recasts and copies tend to be softer and less detailed than on mirrors of earlier periods. Mirrors cast during the Song dynasty tend to have lower tin and higher lead contents than earlier mirrors. All this information must be taken together in order to determine if a mirror was recast.

NOTES ON THE ANALYTICAL DATA

Semiquantitative analyses of the mirrors were performed using a Kevex energy-dispersive x-ray fluorescence spectrometer (Model 075A), with excitation conditions of 50 kV and 3.3 mA using a secondary target of barium and strontium carbonates. A 2-mm collimator on the x-ray tube and a 6-mm collimator on the detector allowed a roughly 4-mm area of each mirror to be analyzed. Data were collected for 150 seconds. At least two sites—one on each side of the mirror—were analyzed. Kevex's "Exact" program, a fundamental parameters equation, was used to calculate the weight percentage of each alloy element. The mass balance of each analysis was generally short of 100 percent due to corrosion, which often preferentially removes copper from the alloy during burial. The results of the analyses were normalized to copper.

The semiquantitative results have been tabulated and reported in chart form for tin, lead, zinc, antimony, arsenic, silver, nickel, and iron. The major components—usually tin and lead—are reported by ranges of 5 percent: >1 indicates that the concentration of an element is between 1 and 4.9 percent; >5 indicates that the concentration is between 5 and 9.9 percent; >10 indicates that the concentration is between 10 and 14.9 percent, and so forth. A "P" indicates that the element was detected and the concentration is between 0.1 and 0.9 percent. "ND" indicates that the element was not detected. "Back" refers to the decorated side, and "Front" to the reflective surface. In the List of the Carter Collection in this Appendix, designations of low, medium, and high have been used to describe each alloy's tin and lead components. For both tin and lead, the "low" designation is for alloys with less than 10 percent while "high" designates alloys with greater than 15 percent. "Medium" describes a content between 10 and 15 percent.

While analysis using x-ray fluorescence spectroscopy is far from ideal, since accurate analyses of corroded mirror surfaces and leaded alloys are problematic, there are also advantages. The technique is not destructive, requiring no sample from the object. It also provides relatively quick alloy assessment. During the analyses mercury occasionally was detected, which may indicate the use of *xuanxi*, a polishing compound containing mercury and tin used on the surface. When mercury was found, its presence is noted in the semiquantitative data.

Five of the mirrors had been broken and repaired, then extensively overpainted to hide the damage. The paint layers usually dramatically altered the analytical results, as paint absorbs x-ray signals from the bronze alloy so that its composition is understated. Conversely, x-ray signals from the paint layer's elements are overstated. In one case the knob of mirror no. 44 has an iron content of 19.8 percent, indicating an iron-bearing pigment was used in the paint. The painted mirrors are also noted.

B. C.

1. Zhu/He 1993, 51.

2. The mirrors were x-rayed at XRI Inc. in Independence, Ohio, using a 1 million eV Linatron manufactured by Varian Corporation.

3. Chase 1983, 109–10.

4. Zhu/He, 52; they illustrate two clay molds, one dating to the Warring States period and the second to the Western Han.

5. Meyers/Holmes 1983, 124–36.

6. Zhu/He, 52.

7. Ibid., 57–58.

8. Ibid., 53

9. Ibid., 53–62.

10. Taube et al.1995, 215–21; Chase/Franklin 1979, 216–41.

11. The x-ray fluorescence equipment cannot detect elements with atomic weights lighter than potassium (nos. 1–18 in the periodic table). The equipment is configured on a tripod so that analyses can be done from the surface of the work of art. This requires that the analyses be done in open air and not in a vacuum or a helium atmosphere. As a result, the air absorbs the x-rays from the lighter-weight elements.

12. Meeks 1993, 81.

13. Bulling/Drew 1971–72, 36–57.

Acc. No.	Location	Date	Sn	Pb	Zn	Sb	As	Ag	Ni	Fe	Hg	Notes
1995.267	Front	Warring States	>15	>20	P	P	P	P	P	P	Yes	
1995.267	Back	Warring States	>20	>15	P	P	P	P	P	P	Yes	
1995.268	Front	Warring States	>25	>5	P	P	P	P	P	P	No	
1995.268	Back	Warring States	>25	>5	P	P	P	P	P	P	No	
1995.269	Front	Warring States	>25	>1	P	P	P	P	ND	P	Yes	
1995.269	Back	Warring States	>25	>10	P	P	P	P	P	P	Yes	
1995.270	Back	Warring States	>25	>5	P	P	P	P	P	P	No	
1995.270	Back, 2nd	Warring States	>25	>5	P	P	P	P	P	P	No	
1995.271	Front	Warring States	>25	>1	P	P	P	P	P	P	Yes	
1995.271	Back	Warring States	>25	>1	P	P	P	P	P	P	Yes	
1995.272	Front	Warring States	>15	>1	P	P	P	P	P	P	No	
1995.272	Back	Warring States	>15	>1	P	P	P	P	P	P	No	
1995.273	Front	Warring States	>20	>1	P	P	P	P	P	P	No	
1995.273	Back	Warring States	>25	>1	P	P	P	P	P	P	No	
1995.274	Front	Western Han	>1	>20	P	P	P	P	P	P	No	
1995.274	Back	Western Han	>5	>20	P	P	P	P	P	P	No	
1995.275	Back	Western Han	>15	>1	P	P	P	P	P	P	No	
1995.275	Back, 2nd	Western Han	>15	>1	P	P	P	P	P	P	No	
1995.276	Front	Western Han	>20	>1	P	P	P	P	P	P	No	
1995.276	Back	Western Han	>20	>1	P	P	P	P	P	P	No	
1995.277	Front, corrosion	Warring States	>20	>1	P	P	P	P	ND	P	No	
1995.277	Front, glossy area	Warring States	>15	>1	P	P	P	ND	P	P	No	
1995.277	Back	Warring States	>15	>1	P	P	ND	ND	P	P	No	
1995.278	Front	Warring States	>25	>10	P	P	P	P	P	P	No	
1995.278	Front, 2nd	Warring States	>15	>5	P	P	P	P	P	P	No	
1995.278	Front, 3rd	Warring States	>15	>5	P	P	P	P	P	P	No	
1995.278	Back	Warring States	>15	>1	P	P	P	P	P	P	No	
1995.278	Back, 2nd	Warring States	>20	>5	P	P	P	P	P	P	No	
1995.278	Back, 3rd	Warring States	>15	>1	P	P	P	P	P	P	No	
1995.279	Front	Warring States	>20	>1	P	P	P	P	P	P	No	
1995.279	Back	Warring States	>25	>1	P	P	P	P	P	P	No	
1995.280	Back	Warring States	>15	>1	P	P	P	P	P	P	No	
1995.280	Back, 2nd	Warring States	>15	>1	P	P	P	P	P	P	No	
1995.281	Front	Warring States	>25	>1	P	P	P	ND	P	P	No	
1995.281	Back	Warring States	>25	>1	P	P	P	P	P	P	No	
1995.282	Front	Warring States	>25	>10	P	P	P	P	P	P	Yes	
1995.282	Back	Warring States	>25	>5	P	P	P	P	P	P	Yes	
1995.283	Front	Western Han	>25	>1	P	P	P	P	P	P	Yes	
1995.283	Back	Western Han	>25	>1	P	P	P	P	P	P	No	
1995.284	Front	Western Han	>25	>5	P	P	P	P	P	P	No	
1995.284	Back	Western Han	>20	>1	P	P	P	P	P	P	No	
1995.285	Front	Western Han	>20	>1	P	P	P	P	P	P	No	
1995.285	Back	Western Han	>20	>1	P	P	P	P	P	P	No	
1995.286	Front	Eastern Han	>20	>1	P	P	P	P	P	P	No	
1995.286	Back	Eastern Han	>20	>1	P	P	P	P	P	P	No	
1995.287	Front	Western Han	>25	>1	P	P	P	P	P	P	No	
1995.287	Back	Western Han	>25	>5	P	P	P	P	P	P	No	
1995.288	Front	Western Han	>25	>1	>1	>1	>1	P	P	P	No	
1995.288	Back	Western Han	>20	>1	P	P	P	P	P	P	No	
1995.289	Front	Western Han	>25	>1	P	P	P	P	P	P	No	
1995.289	Back	Western Han	>25	>1	P	P	P	P	P	P	No	
1995.290	Front	Western Han	>20	>1	P	P	P	P	P	P	No	
1995.290	Back	Western Han	>20	P	P	P	P	P	P	P	No	

Acc. No.	Location	Date	Sn	Pb	Zn	Sb	As	Ag	Ni	Fe	Hg	Notes
1995.291	Front	Western Han	>25	>1	P	P	P	P	P	P	No	
1995.291	Back	Western Han	>25	>1	P	P	P	P	P	P	No	
1995.292	Front	Western Han	>25	>5	P	P	P	P	P	P	No	
1995.292	Back	Western Han	>25	>1	P	P	P	P	P	P	Yes	
1995.293	Front	Japan/Edo	>20	>5	P	P	P	P	ND	P	Yes	
1995.293	Front, 2nd	Japan/Edo	>20	>5	P	P	P	P	ND	P	Yes	
1995.293	Front, 3rd	Japan/Edo	>20	>5	P	P	P	P	P	P	Yes	
1995.293	Front, 4th	Japan/Edo	>15	>10	P	P	P	P	P	P	Yes	
1995.293	Back	Japan/Edo	>15	>20	P	P	P	P	P	P	Yes	
1995.293	Back, 2nd	Japan/Edo	>20	>5	P	P	P	P	ND	P	Yes	
1995.293	Back, 3rd	Japan/Edo	>15	>20	P	P	P	P	P	P	Yes	
1995.293	Back, 4th	Japan/Edo	>15	>15	P	P	P	P	P	P	Yes	
1995.293	Back, 5th	Japan/Edo	>15	>20	P	P	P	P	P	P	Yes	
1995.294	Front	Western Han	>20	>1	P	P	P	P	P	P	No	
1995.294	Back	Western Han	>20	>1	P	P	P	P	P	P	No	
1995.295	Front	Xin	>25	>1	P	P	P	P	P	P	No	
1995.295	Back	Xin	>25	>1	P	P	P	P	P	P	No	
1995.296	Front	Eastern Han	>25	>1	P	P	P	P	ND	P	No	
1995.296	Back	Eastern Han	>25	>1	P	P	P	P	P	P	No	Paint
1995.296	Front, cent.	Eastern Han	>20	>1	P	P	P	P	ND	P	No	
1995.296	Knob	Eastern Han	>25	>5	P	P	P	P	P	P	No	Paint
1995.297	Front	Western Han	>25	>1	P	P	P	P	P	P	Yes	
1995.297	Back	Western Han	>25	>5	P	P	P	P	P	P	No?	
1995.298	Front	Xin	>20	>1	P	P	P	P	P	P	No	
1995.298	Back	Xin	>20	>1	P	P	P	P	P	P	No	
1995.299	Front	Xin	>25	>1	P	P	P	P	P	P	Yes	
1995.299	Back	Xin	>25	>1	P	P	P	P	P	P	Yes	
1995.300	Front	Western Han	>20	>1	P	P	P	P	P	P	No	
1995.300	Back	Western Han	>20	>1	P	P	P	P	P	P	No	
1995.301	Front	Western Han	>20	>1	P	P	P	P	P	P	No	
1995.301	Back	Western Han	>25	>1	P	P	P	P	P	P	No	
1995.302	Front	Western Han	>25	>5	P	P	P	P	P	P	No	
1995.302	Knob	Western Han	>25	>5	P	P	P	P	P	P	No	
1995.303	Front	Western Han	>25	>1	P	P	P	P	P	P	No	
1995.303	Back	Western Han	>25	>1	P	P	P	P	P	P	No	
1995.304	Front	Eastern Han	>25	>1	P	P	P	ND	ND	P	Yes	
1995.304	Front, 2nd	Eastern Han	>25	>1	P	P	P	P	P	P	No	
1995.304	Knob	Eastern Han	>25	>1	P	P	P	P	ND	P	No	
1995.305	Front	Western Han	>25	>1	P	P	P	P	P	P	Yes	
1995.305	Back	Western Han	>25	>1	P	P	P	P	ND	P	Yes	
1995.306	Front	Western Han	>20	>1	P	P	P	P	P	P	No	
1995.306	Back	Western Han	>25	>1	P	P	P	P	P	P	No	
1995.307	Front	Western Han	>25	>1	P	P	P	P	P	P	No	
1995.307	Back	Western Han	>20	>1	P	P	P	P	P	P	No	
1995.308	Front	Western Han	>25	>1	P	P	P	P	ND	P	No	
1995.308	Back	Western Han	>25	>1	P	P	P	P	ND	P	Yes	
1995.309	Front	Western Han	>20	>1	P	P	P	P	P	P	No	
1995.309	Back	Western Han	>25	>1	P	P	P	P	P	P	No	
1995.310	Front	Western Han	>20	>5	P	P	P	P	P	P	No	
1995.310	Back	Western Han	>20	>1	P	P	P	P	P	P	No	
1995.311	Front	Western Han	>20	>1	P	P	P	P	P	P	No	
1995.311	Back	Western Han	>20	>1	P	P	P	P	P	P	No	
1995.312	Front	Western Han	>5	>10	P	P	P	P	P	P	No	Corrosion

Acc. No.	Location	Date	Sn	Pb	Zn	Sb	As	Ag	Ni	Fe	Hg	Notes
1995.312	Back	Western Han	>20	>1	P	P	P	P	P	P	No	
1995.312	Back, 2nd	Western Han	>20	>1	P	P	P	P	P	P	No	
1995.313	Front	Western Han	>25	>5	P	>1	P	P	P	P	No	
1995.313	Back	Western Han	>25	>5	P	P	P	P	P	P	No	
1995.314	Front	Western Han	>25	>1	P	P	P	P	P	P	No	
1995.314	Back	Western Han	>25	>1	P	P	P	P	P	P	No	
1995.315	Front	Eastern Han/Jin	>25	>1	P	P	P	P	P	P	No	
1995.315	Back	Eastern Han/Jin	>25	>1	P	P	P	P	P	P	No	
1995.316	Front	Eastern Han/Jin	>20	>1	P	P	P	P	P	P	No	
1995.316	Back	Eastern Han/Jin	>20	>1	P	P	P	P	P	P	No	
1995.317	Front	Three King./W. Jin	>25	>1	P	P	P	P	P	P	No	
1995.317	Back	Three King./W. Jin	>25	>1	P	P	P	P	ND	P	No	
1995.318	Front	Eastern Han	>20	>1	P	P	P	P	P	P	No	
1995.318	Back	Eastern Han	>20	>1	P	P	P	P	P	P	No	
1995.319	Front	Eastern Han	>20	>5	P	P	P	P	P	P	No	
1995.319	Back	Eastern Han	>20	>1	P	P	P	P	P	P	No	
1995.320	Front	Eastern Han	>25	>1	P	P	P	P	P	P	No	
1995.320	Knob	Eastern Han	>25	>1	P	P	P	P	P	P	No	
1995.321	Front	Eastern Han	>20	>1	P	P	P	P	P	P	No	
1995.321	Back	Eastern Han	>25	>1	P	P	P	P	P	P	No	
1995.322	Front	Three Kingdoms	>25	>1	P	P	P	P	P	P	No	
1995.322	Knob	Three Kingdoms	>25	>1	P	P	P	P	P	P	No	
1995.323	Front	Three Kingdoms	>25	>1	P	P	P	P	P	P	No	
1995.323	Knob	Three Kingdoms	>25	>1	P	P	P	P	P	P	No	
1995.324	Front	Eastern Han	>25	>1	P	P	P	P	P	P	No	
1995.324	Knob	Eastern Han	>25	>1	P	P	P	P	P	P	No	
1995.325	Front	Eastern Han	>5	>5	P	P	P	>1	P	P	No	
1995.325	Back	Eastern Han	>1	>20	P	P	P	ND	P	P	No	
1995.325	Knob	Eastern Han	>5	>20	P	P	P	P	P	P	No	
1995.326	Front	Eastern Han	>25	>1	P	P	P	P	P	P	No	
1995.326	Knob	Eastern Han	>25	>1	P	P	P	P	P	P	No	
1995.327	Front	Six Dynasties	>25	>1	P	P	P	P	P	P	No	
1995.327	Back	Six Dynasties	>25	>1	P	P	P	P	ND	P	No	
1995.328	Front	Eastern Han	>20	>1	P	P	P	P	P	P	No	
1995.328	Back	Eastern Han	>20	>1	P	P	P	P	P	P	No	
1995.329	Front	Eastern Han	>25	>1	P	P	P	P	P	P	No	
1995.329	Back	Eastern Han	>25	>1	P	P	P	P	P	P	No	
1995.330	Front	Eastern Han	>25	P	P	P	P	P	P	P	No	
1995.330	Back	Eastern Han	>25	>1	P	P	P	P	P	P	Yes	
1995.330	Knob	Eastern Han	>25	>1	P	P	P	P	P	P	No	Paint
1995.331	Front	Eastern Han	>25	P	ND	P	ND	P	P	P	No	Paint
1995.331	Knob	Eastern Han	>25	P	P	P	P	P	P	P	No	Paint
1995.332	Front	Eastern Han	>25	P	P	P	P	P	P	P	Yes	
1995.332	Knob	Eastern Han	>25	>1	P	P	P	P	P	P	No	Paint
1995.333	Front	Eastern Han	>25	>1	P	P	P	P	P	P	No	
1995.333	Knob	Eastern Han	>25	>1	P	P	P	P	P	P	No	
1995.334	Front	Western Jin	>20	>1	P	P	ND	ND	P	P	No	Paint
1995.334	Knob	Western Jin	>15	>1	P	P	P	ND	P	P	No	Paint
1995.335	Front	Tang	>20	>1	P	P	P	P	P	P	No	
1995.335	Knob	Tang	>20	>1	P	P	P	P	P	P	No	
1995.336	Front	Sui	>15	>5	P	P	P	P	P	P	Yes	
1995.336	Back	Sui	>20	>10	P	P	P	P	P	P	Yes	
1995.337	Front	Sui	>20	>1	P	P	P	ND	P	P	No	

Acc. No.	Location	Date	Sn	Pb	Zn	Sb	As	Ag	Ni	Fe	Hg	Notes
1995.337	Knob	Sui	>15	>1	P	P	P	P	P	P	No	
1995.338	Front	Tang	>20	>1	P	P	P	P	P	P	No	
1995.338	Knob	Tang	>20	>1	P	P	P	P	P	P	No	
1995.339	Front	Sui	>20	P	P	P	P	ND	P	P	No	
1995.339	Knob	Sui	>20	>1	P	P	ND	P	P	P	No	
1995.340	Front	Tang	>20	>1	P	P	P	P	ND	P	No	
1995.340	Back	Tang	>20	>1	P	P	P	ND	ND	P	No	
1995.341	Front	Tang	>20	>1	P	P	P	P	P	P	No	
1995.341	Back	Tang	>20	>1	P	P	P	P	P	P	No	
1995.342	Front	Tang	>25	>5	P	P	P	P	P	P	No	
1995.342	Back	Tang	>20	>1	P	P	P	P	P	P	No	
1995.343	Front	Tang	>20	>1	P	P	P	P	P	P	Yes	
1995.343	Back	Tang	>20	>1	P	P	P	P	P	P	Yes	
1995.344	Front	Tang	>20	>1	P	P	P	P	P	P	No	
1995.344	Knob	Tang	>20	>1	P	P	P	P	P	P	No	
1995.345	Front	Tang	>20	>1	>1	>1	>1	P	P	P	Yes	
1995.345	Knob	Tang	>20	P	>1	>1	>1	P	P	P	Yes	
1995.346	Front	Tang	>20	>1	P	P	P	P	P	P	No	
1995.346	Knob	Tang	>20	>1	P	P	P	P	P	P	No	
1995.347	Front	Tang	>20	>1	P	P	P	P	P	P	No	
1995.347	Knob	Tang	>20	>1	P	P	P	P	P	P	No	
1995.348	Front	Tang	>25	>1	P	P	P	P	P	P	No	
1995.348	Knob	Tang	>25	>1	P	P	P	P	P	P	No	
1995.349	Front	Tang	>20	>1	ND	ND	ND	ND	P	P	No	
1995.349	Back	Tang	>20	>1	P	P	P	P	P	P	No	
1995.350	Front	Tang	>20	>1	P	P	P	P	P	P	No	
1995.350	Back	Tang	>20	>1	P	P	P	P	P	P	No	
1995.351	Front	Tang	>20	>10	P	P	P	P	P	P	No	
1995.351	Knob	Tang	>20	>1	P	P	P	P	P	P	No	
1995.352	Front	Tang	>15	>1	>1	>1	>1	P	P	P	No	
1995.352	Back	Tang	>20	>1	>1	>1	>1	P	P	P	No	
1995.353	Front	Tang	>20	P	P	P	P	P	P	P	No	
1995.353	Knob	Tang	>20	>1	P	P	P	ND	P	P	No	
1995.354	Front	Tang	>25	P	P	P	P	P	P	P	No	
1995.354	Back	Tang	>20	>1	P	P	P	P	P	P	Yes	
1995.355	Back	Tang	>15	>1	>1	>1	>1	P	P	P	Yes	
1995.355	Edge	Tang	>15	>1	>1	>1	>1	P	P	P	No	
1995.356	Front	Tang	>25	>1	P	P	P	P	P	P	Yes	
1995.356	Knob	Tang	>25	>1	P	P	P	P	P	P	Yes	
1995.357	Front	Jin	>25	>5	P	P	P	P	P	P	No	
1995.357	Knob	Jin	>20	>1	P	P	P	P	P	P	No	
1995.358	Front	Tang	>10	>1	>15	>1	>1	P	P	P	No	
1995.358	Knob	Tang	>10	>1	>15	>1	>1	P	P	P	No	
1995.359	Front	Tang	>20	P	P	P	P	P	P	P	No	
1995.359	Knob	Tang	>20	P	P	P	P	P	P	P	No	
1995.360	Front	Tang	>20	>1	P	P	P	P	P	P	No	
1995.360	Back	Tang	>20	>1	P	P	P	P	P	P	No	
1995.361	Front	Tang	>20	>1	P	P	P	P	P	P	No	
1995.361	Back	Tang	>20	>1	P	P	P	P	P	P	No	
1995.362	Front	Tang	>20	>1	P	P	P	P	P	P	No	
1995.362	Back	Tang	>20	>1	P	P	P	P	P	P	No	
1995.363	Front	Tang	>20	>1	P	P	P	P	P	P	No	
1995.363	Knob	Tang	>20	P	P	P	P	P	P	P	No	

Acc. No.	Location	Date	Sn	Pb	Zn	Sb	As	Ag	Ni	Fe	Hg	Notes
1995.364	Front	Tang	>20	>1	P	P	P	P	P	P	No	
1995.364	Knob	Tang	>20	>1	P	P	P	P	P	P	No	
1995.365	Front	Tang	>20	>1	P	P	P	ND	P	P	No	
1995.365	Knob	Tang	>20	>5	P	P	ND	ND	ND	P	No	
1995.366	Front	Tang	>15	>1	P	P	P	P	P	P	No	
1995.366	Back	Tang	>15	>1	P	P	P	ND	ND	P	No	
1995.367	Front	Tang	>25	>1	P	P	P	P	P	P	No	
1995.367	Back	Tang	>25	>1	P	P	P	P	P	P	No	
1995.368	Front	Liao	>15	>1	P	P	P	P	P	P	No	
1995.368	Back	Liao	>20	>1	P	P	P	P	P	P	No	
1995.369	Front	Tang	>20	>1	P	P	P	P	P	P	No	
1995.369	Back	Tang	>20	>1	P	P	P	P	P	P	No	
1995.370	Front	Ming	>10	>1	>1	>1	>1	P	P	P	No	
1995.370	Back	Ming	>15	>1	>10	>1	>1	P	P	P	No	
1995.370	Back, 2nd	Ming	>15	>1	>10	>1	>1	P	P	P	Yes	
1995.371	Front	Song	>15	>1	P	P	P	P	P	P	Yes	
1995.371	Back	Song	>15	>1	P	P	P	P	P	P	Yes	
1995.372	Front	Song	>20	>10	P	P	P	P	P	P	No	
1995.372	Back	Song	>25	>5	P	P	P	P	ND	P	No	
1995.373	Front	Tang	>15	>1	P	P	P	P	P	P	No	
1995.373	Back	Tang	>15	>1	P	P	P	P	P	P	No	
1995.374	Front	Song	>25	>1	P	P	P	P	P	P	No	
1995.374	Front, 2nd	Song	>25	>1	P	P	P	P	P	P	No	
1995.374	Knob	Song	>20	>1	P	P	P	P	P	P	No	
1995.375	Front	Jin	>25	>1	P	P	P	>1	ND	P	No	
1995.375	Knob	Jin	>25	>1	P	P	P	>1	P	P	No	
1995.376	Front	Song/Jin	>25	>5	P	P	P	P	ND	P	No	
1995.376	Knob	Song/Jin	>25	>5	P	P	P	P	P	P	Yes	
1995.377	Knob	Ming	>20	>5	P	P	P	P	P	P	No	
1995.378	Knob	Song	>5	>20	P	P	P	P	P	P	No	
1995.379	Back	Jin	>10	>20	P	P	P	P	P	P	No	
1995.379	Knob	Jin	>10	>20	P	P	P	P	ND	P	No	
1995.380	Front	N. Song	>15	>1	P	P	P	P	P	P	No	
1995.380	Back	N. Song	>20	>1	>1	>1	P	P	P	P	No	
1995.381	Front	Jin	>5	>20	P	P	P	P	P	P	No	
1995.381	Back	Jin	>5	>20	P	P	P	P	P	P	Yes (small)	
1995.382	Front	Jin	>5	>20	P	P	P	P	P	P	Yes	
1995.382	Front, 2nd	Jin	>5	>20	P	P	P	P	P	P	Yes	
1995.382	Knob	Jin	>15	>20	P	P	P	P	P	P	No	
1995.383	Knob	Song	>15	>20	P	P	P	P	P	P	No	
1995.384	Front	Jin/Yuan	>10	>20	P	P	P	P	ND	P	No	
1995.384	Back	Jin/Yuan	>10	>20	P	P	P	P	ND	P	No	
1995.385	PR Side	Qing	>5	>1	P	P	P	P	P	P	No	
1995.385	PL Neck	Qing	>5	>5	P	P	P	P	P	P	No	
1995.385	Bracket	Qing	>5	>5	P	P	P	P	P	P	No	
1995.386	PR Side	Qing	>1	>1	P	P	ND	ND	P	P	No	
1995.386	PL Side	Qing	>1	>5	P	P	ND	P	P	P	No	
1995.387	Front	Jin	>20	>5	P	P	P	P	P	P	No	
1995.387	Knob	Jin	>25	>10	P	P	P	P	P	P	No	
1995.387	Knob, 2nd	Jin	>25	>10	P	P	P	P	P	P	No	
1995.388	Front	Yuan	>5	>10	P	P	P	P	P	P	No?	
1995.388	Knob	Yuan	>5	>15	P	P	P	P	P	P	Yes	
1995.389	Front	Song	>15	>10	P	P	P	P	P	P	No	

Acc. No.	Location	Date	Sn	Pb	Zn	Sb	As	Ag	Ni	Fe	Hg	Notes
1995.389	Knob	Song	>15	>15	P	P	P	P	P	P	Yes	
1995.389	Knob, 2nd	Song	>15	>20	P	P	P	P	P	P	No	
1995.390	Front	Unknown	>1	>1	P	P	P	ND	P	P	No	
1995.390	Back	Unknown	>1	>1	P	P	P	ND	P	P	No	
1995.391	Front	Yuan	>15	>5	P	P	P	P	P	P	Yes	
1995.391	Back	Yuan	>15	>10	P	P	P	P	P	P	Yes	
1995.392	Front	Jin	>10	>1	>1	>1	>1	P	P	P	No	
1995.392	Knob	Jin	>10	>1	>1	>1	>1	P	P	P	Yes	
1995.393	Front	Ming	>15	>5	P	P	ND	P	P	P	No	
1995.393	Knob	Ming	>15	>10	P	P	ND	P	P	P	No	
1995.393	Knob, 2nd	Ming	>15	>10	P	P	P	P	P	P	Yes	
1995.394	Front	Five Dynasties	>20	>5	P	P	P	>1	P	P	No	
1995.394	Back	Five Dynasties	>20	>10	P	P	P	P	P	P	Yes	
1995.395	Front	Ming	>10	>5	P	P	P	P	P	P	No	
1995.395	Back	Ming	>10	>5	P	P	P	P	P	P	Yes	
1995.396	Front	Japan/Edo	>20	>1	P	P	P	P	P	P	No	
1995.396	Knob	Japan/Edo	>20	P	P	P	P	P	P	P	No	
1995.397	Front	Japan/Edo	>1	>1	P	P	P	P	P	P	No	
1995.397	Knob	Japan/Edo	>1	P	ND	ND	ND	P	P	P	Yes	
1995.398	Front	Japan/Edo	ND	P	P	P	ND	P	ND	P	No?	
1995.398	Knob	Japan/Edo	ND	P	ND	P	ND	P	P	P	No?	
1995.399	Back	Japan/Edo	ND	>1	>20	>5	>1	P	P	P	No	
1995.399	Front, handle	Japan/Edo	P	>1	>20	>5	>1	P	P	P	No	
1995.399	Back, handle	Japan/Edo	ND	>1	>20	>5	>1	P	P	P	No	
1995.400	Front	Japan/Edo	>5	>1	P	P	P	ND	P	P	No	
1995.400	Back	Japan/Edo	>5	>1	P	P	P	P	P	P	Yes	
1995.401	Front	Qing	>1	>1	P	P	P	ND	P	P	No?	
1995.402	Front	Near East	P	P	P	P	P	ND	P	P	No?	
1995.402	Back	Near East	P	P	P	P	P	ND	P	P	No	
1995.403	Front	Near East	>25	>1	P	P	P	P	P	P	No	
1995.403	Back	Near East	>20	P	P	P	P	P	P	P	No	
1999.217	Front	Tang	>20	>1	>1	>1	P	P	P	P	No	
1999.217	Front, 2nd	Tang	>15	>1	P	P	P	P	P	P	No	
1999.218	Front	Tang	>20	>1	P	P	P	P	P	P	Yes	
1999.218	Back	Tang	>20	>1	P	P	P	P	P	P	Yes	
1999.219	Front	Sui	>20	>1	P	P	P	P	P	P	No	
1999.219	Back	Sui	>20	>1	P	P	P	P	P	P	No	
1999.220	Front	Warring States	>15	>5	P	P	P	P	P	P	No	
1999.220	Back	Warring States	>5	>20	P	P	ND	P	P	P	No	
1999.221	Front	Yuan	>5	>20	P	P	ND	P	P	P	No	
1999.221	Back	Yuan	>5	>20	P	P	ND	P	P	P	Yes	
1999.222	Front	Song	>15	>1	P	P	P	P	P	P	Yes	
1999.222	Back	Song	>15	>1	P	P	P	P	P	P	Yes	

List of the Carter Collection

Chinese Mirrors and Stands

四葉羽紋鏡 (no. 1)
Mirror with Quatrefoil and Feather Pattern
Warring States (475–221 BC)
Weight: 90.5 g
Diameter: 11.2 cm
Rim height: 0.35 cm
Overall height (knob): 0.6 cm
1995.267

Manufacture: Cast in high-tin, high-lead bronze in a clay mold. Abrasion marks on the concave rim indicate this part of the back was finished. Little finishing work is evident in the relief and background areas, but thick layers of corrosion make determination difficult.

Condition: Thick green, blue, black, and red corrosion products cover nearly all surfaces of the mirror on both sides. One or two faint areas of plain-weave textile pseudomorphs are present on the reflective surface, along with some burial mud. A Chinese character in black ink appears on the corrosion products on the reflective surface.

蟠螭鏡
Mirror with Serpentine Interlaces
Warring States (475–221 BC)
Weight: 43 g
Diameter: 8.1 cm
Rim height: 0.25 cm
Overall height (knob): 0.5 cm
1995.268

Manufacture: Cast in high-tin, low-lead bronze in a clay mold. Some surface porosity can be seen on the back and scattered porosity is visible in the x-radiograph. There are abrasive marks on the back. The concave edge band and center bands were abraded to smooth the metal. The high points of the cast decoration are also abraded. No finishing work was done to the recessed areas.

Condition: Both sides of the mirror have a glossy black surface, with small patches of green and blue corrosion products. The reflective surface has a fine network of surface cracks.

蟠螭鏡
Mirror with Serpentine Interlaces
Warring States (475–221 BC)
Weight: 374 g
Diameter: 18.2 cm
Rim height: 0.75 cm
Overall height (knob): 1 cm
1995.269

Manufacture: Cast in high-tin, low-lead bronze in a clay mold. X-radiographs reveal large areas of porosity. The degree of finishing work is somewhat difficult to determine because of the coating applied to hide the breaks. Some finishing work in the form of rough abrasion is present on the concave border area and in the concave circle near the center.

Condition: X-radiographs reveal that the mirror was broken into four pieces. They also show two blind cracks. The reflective surface was painted to hide repairs. Much of the back is covered with a brown wax-like material that incorporates sand or other grit. A glossy black patina is visible through gaps in the brown coating on the mirror back. Other corrosion products are not evident.

蟠螭菱紋鏡 (no. 5)
Mirror with Serpentine Interlaces and Angular Meanders
Warring States (475–221 BC)
Weight: 258 g
Diameter: 16.9 cm
Rim height: 0.5 cm
Overall height (knob): 1 cm
1995.270

Manufacture: Cast in high-tin, low-lead bronze in a clay mold. A ridge of bronze indicates where the mold cracked on the outermost recessed band in an area relatively free of green corrosion. Abrasive polishing smoothed the raised areas of the decoration. The extent of finishing work in low areas cannot be determined due to thick corrosion layers.

Condition: Two blind cracks extend inward from the edge nearly opposite each other; one is 4 cm long, the other is 3 cm long. Four pieces broke off the edge and were reattached. In the x-radiographs, the pieces appear to be tacked in place with a lead-based solder, indicated by the degree of radio-opacity. Losses were filled with a less radio-opaque material, except for one that was filled with the lead-based solder. The repairs were disguised using ground-up corrosion products and by applying burial mud. Both sides are covered with thick layers of green and red corrosion products. Two small areas of metal on the mirror back are corrosion free. Much burial dirt is trapped in the corrosion and recessed areas. A plain-weave textile pseudomorph is present on the reflective surface at the edge.

蟠螭菱紋鏡
Mirror with Serpentine Interlaces and Angular Meanders
Warring States (475–221 BC)
Weight: 97.5 g
Diameter: 11.7 cm
Rim height: 0.45 cm
Overall height (knob): 0.6 cm
1995.271

Manufacture: Cast in high-tin, low-lead bronze in a clay mold. The raised areas of the back were finished by abrasion. Little to no finishing work is seen in the background around the raised decoration.

Condition: The mirror was broken into three sections, which were then joined by tacking with lead-based solder and with a less dense material such as glue. The joints were filled with putty and inpainted. A thin, compact green and black corrosion layer covers both sides of the mirror's surface. In several areas the corrosion process has etched the metal's dendritic structure. Some shallow, post-burial abrasion is present on the reflective surface. At least three blind cracks emanate from the mirror's edge, which also shows several small nicks and dents.

蟠螭菱紋鏡
Mirror with Serpentine Interlaces and Angular Meanders
Warring States (475–221 BC)
Weight: 247 g
Diameter: 17.9 cm
Rim height: 0.2 cm
Overall height (knob): 0.9 cm
1995.272

Manufacture: Cast in high-tin, low-lead bronze in a clay mold. Some finishing work is evident on the back. The concave bands at the edge and near the center were smoothed and polished. No finishing work is seen on the center knob. Rough abrasion marks are present on the raised relief decoration. Little work is evident in the recessed areas. Cracks from the mold appear as ridges in the cast bronze.

Condition: Much of the silver-colored metal is well preserved. Thick green and red corrosion products cover approximately 15% of the back. Thinner black corrosion products are also present. Approximately 50% of the reflective surface is covered with green, red, and black corrosion products. Plain-weave textile pseudomorphs are preserved in the green corrosion layers on each side.

蟠虺連弧紋鏡
Mirror with Continuous Arcs and Quasi-Dragons
Western Han (206 BC–AD 8)
Weight: 38 g
Diameter: 7.7 cm
Rim height: 0.2 cm
Overall height (knob): 0.4 cm
1995.273

Manufacture: Cast in high-tin, low-lead bronze in a clay mold. The raised flat areas and concave border band of the design were finished by grinding and polishing, leaving abrasion marks in the metal, while the recessed areas have a rough "as-cast" appearance.

Condition: Patchy thin black-colored corrosion products appear on both sides of the mirror. The green and red deposits tend to be thicker on the back. Numerous deep random scratches are in the green corrosion layers on the reflective surface.

蟠虺連弧紋鏡
Mirror with Continuous Arcs and Quasi-Dragons
Western Han (206 BC–AD 8)
Weight: 53 g
Diameter: 18.5 cm
Rim height: 0.25 cm
Overall height (knob): 0.6 cm
1995.274

Manufacture: Cast in low-tin, high-lead bronze in a clay mold. The scalloped decoration and concave border were finished by abrasion and polishing. The extent of additional finishing work is difficult to determine due to the thick, extensive corrosion.

Condition: Corrosion products cover both sides of the mirror. Red and green are found on the reflective surface while green, black, and red are present on the back. On both sides, white accretions mix with the corrosion products and plain-weave textile impressions appear in the corrosion layers. Several recent scratches show on the concave border. A deep nick at the edge reveals a copper-colored metal. The mirror was bent slightly in the area of the nick.

方格四乳鏡 (no. 10)
Mirror with a Square Band and Four Nipples
Western Han (206 BC–AD 8)
Weight: 62 g
Diameter: 8.4 cm
Rim height: 0.35 cm
Overall height (knob): 0.7 cm
Inscription: 見日之光，長樂未央。
1995.275

Manufacture: Cast in high-tin, low-lead bronze in a clay mold. Surface porosity appears on the back. A crack in the mold is seen as a ridge in the cast metal. Some damage to the mold is visible in the inscription square. Finishing work is evident on the back. The concave channel forming the square around the center knob was chiseled. No finishing work is seen on the four nipples. The recessed areas were not finished, as the metal has a rough-cast surface. The concave edge band has abrasion marks from the polishing work.

Condition: Extensive thick red and green corrosion products cover the entire surface. Corrosion products near the edge preserve plain-weave textile pseudomorphs and some textile fibers. Much silver-colored metal is preserved on the back, with approximately one-third of the back covered with green, red, and black corrosion deposits, some of them thick.

四花連弧鏡 (no. 12)
Mirror with Quatrefoils and Linked Arcs
Western Han (206 BC–AD 8)
Weight: 234.5 g
Diameter: 14 cm
Rim height: 0.6 cm
Overall height (knob): 0.6 cm
1995.276

Manufacture: Cast in high-tin, low-lead bronze in a clay mold. Finishing work is evident: raised areas were polished and heart shapes and interlinking channels were chiseled. The recessed areas show little to no finishing work.

Condition: Overall, the mirror has an unusual matte black appearance, perhaps from being in a fire or subjected to high heat after excavation. On the back are areas of accretion with an iridescent quality somewhat similar to glass decomposition products. The reflective surface has a matte, mottled appearance. Other than several small areas of red corrosion scattered on both sides, few corrosion products are present. A deep chip loss at the rim exposes silver-colored bronze.

蟠螭菱紋鏡 (no. 6)
Mirror with Serpentine Interlaces and Angular Meanders
Warring States (475–221 BC)
Weight: 371 g
Diameter: 17 cm
Rim height: 0.7 cm
Overall height (knob): 0.9 cm
1995.277

Manufacture: Cast in high-tin, low-lead bronze in a clay mold. Some rough finishing work is evidenced by abrasive marks on the concave edge and the concave circle near the mirror's center. The flat raised decoration has a rough cast metal surface, with no apparent finishing. Ridges of metal where cracks occurred in the mold were not removed.

Condition: The mirror is extensively corroded. Its back is a mottled green and black, with most of the green found in localized corrosion pits. Some burial mud is present on the back. Much of the reflective surface is coated with a shiny black material that appears to be lacquer, leaving several spots of extensive corrosion exposed. Red and green corrosion products are in these exposed areas. There is a hole in the mirror in the largest corrosion pit. Putty was applied in this area to close up a larger hole, and can also be seen on the back. The lacquer layer probably conceals other deep corrosion pits.

雲雷紋地連弧紋鏡 (no. 9)
Mirror with Continuous Arcs against Whorl Pattern
Warring States (475–221 BC)
Weight: 557 g
Diameter: 23.7 cm
Rim height: 0.75 cm
Overall height (knob): 1.2 cm
1995.278

Manufacture: Cast in high-tin, low-lead bronze in a ceramic mold. Some surface porosity and irregularities are present at the back edge. The flat edge band, scalloped band, and the circle near the center were carefully finished with concave surfaces. Little to no finishing work was done in the gold-tone whorl pattern. X-ray fluorescence analyses did not reveal the presence of gold and there was virtually no difference in composition between the silver and gold areas.

Condition: Three cracks are present—one in the center area near the knob, and two at the edge which were partially filled and inpainted. Extensive thick green and red and thinner black corrosion products cover the reflective surface. Several areas of bright silver-colored metal are seen. Thick green corrosion products partially cover the back, primarily on the edge and polished areas. The whorl field is mostly corrosion free. Some plain-weave textile pseudomorphs are visible on the mirror edge.

雙花龍鳳鏡 (no. 8)
Mirror with Double Quatrefoils, Dragons, and Phoenixes
Warring States (475–221 BC)
Weight: 805 g
Diameter: 23.2 cm
Rim height: 0.5 cm
Overall height (knob): 1.2 cm
1995.279

Manufacture: Cast in high-tin, low-lead bronze in a clay mold. Abrasive finishing work is evident on the rim, flat relief decoration, and four lobe shapes. Some ridges of metal were caused by mold cracks.

Condition: Numerous cracks and breaks indicate that the mirror was broken into nine pieces, which then were joined with a lead-based solder. The joins were skillfully concealed by carving the solder to follow the decoration and then covering it with ground-up copper corrosion products. The reflective surface is extensively covered with thick layers of green and red corrosion products. Textile pseudomorphs are seen throughout. At least two different plain-weave textiles can be identified by the differing thread diameter and tightness of the weave pattern. The back also has thick deposits of green and red corrosion products, but these are scattered over the surface along with some thinner black and gray corrosion. Much of the surface is a dull silver-colored metal. Some white deposits—possibly polish residues—are evident.

三龍鏡 (no. 7)
Mirror with Three Dragons
Warring States (475–221 BC)
Weight: 249 g
Diameter: 16.9 cm
Rim height: 0.45 cm
Overall height (knob): 0.6 cm
1995.280

Manufacture: Cast in high-tin, low-lead bronze in a ceramic mold. The raised flat decoration and concave border were abraded and polished. There is no evidence of chisel work. Little to no finishing work was done in the low areas of the design.

Condition: The mirror was broken at the edge, with two triangular-shaped losses that measure 1.5 cm or less on each side. One piece has been reattached. The losses were filled with a less radio-opaque material—perhaps aluminum. A bright aluminum-colored spot can be observed in one area of the repair. A blind crack is present. The mirror back remains essentially metallic, with a dark silver-gray appearance. Thick green and red corrosion layers cover approximately half of the back. The same type of corrosion products completely cover the reflective surface. On both sides, plain-weave textile pseudomorphs appear in the green corrosion layer. Textile fibers are found on the reflective side near the perimeter and also on the mirror back.

四山鏡 (no. 2)
Mirror with Four T's
Warring States (475–221 BC)
Weight: 124.5 g
Diameter: 10.4 cm
Rim height: 0.45 cm
Overall height (rim): 0.45 cm
1995.281

Manufacture: Cast in high-tin, low-lead bronze in a clay mold. Some porosity appears at the mirror edges and is evident in the x-radiograph. There is extensive finishing work on the border band and in the T-shapes. Abrasive marks are found on the border. Scrape marks left by chisels can be seen on the concave T-shapes and the center square. Little to no finishing work was done on the rest of the mirror back. The cast decoration is rough and irregular on a microscopic level.

Condition: The mirror has a lustrous black patina on both the front and back surfaces. Shallow subsurface warts of corrosion are scattered over the reflective surface, and some green corrosion is present at the edge. A faint impression of a plain-weave textile is visible in the center of the reflective surface. Little corrosion is found on the mirror back. A 4.7-cm-long blind crack starts at the edge and extends to the mirror's center area. On the opposite side is a Y-shaped internal blind crack visible only in the x-radiograph.

五花折疊式菱紋鏡 (no. 3)
Mirror with Five Blossoms with Overlapping Lozenges
Warring States (475–221 BC)
Weight: 105 g
Diameter: 11.4 cm
Rim height: 0.4 cm
Overall height (rim): 0.4 cm
1995.282

Manufacture: Cast in high-tin, low-lead bronze in a clay mold. Chisel marks show in the concave X-shaped channels. The concave border also was abraded to smooth and polish the metal, and some abrasion is evident on the leaf or petal part of the decoration. Several large ridges of metal from mold cracks were left in place—including the concave border, which for the most part was abraded. The extent of finish work in the background is not known due to the thick corrosion.

Condition: The center knob is lost. A network of blind cracks in the center area of the mirror shows on both sides. The back is covered with green, red, and black corrosion products. The reflective surface is glossy but mottled with black and green corrosion products. A red corrosion product resides in pits that have opened up on the surface.

蟠螭紋博局式鏡 (no. 24)
TLV Mirror with Serpentine Interlaces
Western Han (206 BC–AD 8)
Weight: 302.5 g
Diameter: 14 cm
Rim height: 0.5 cm
Overall height (knob): 1 cm
Inscription: 大樂貴富，得所好。千秋萬歲，延年益壽。
1995.283

Manufacture: Cast in high-tin, low-lead bronze in a clay mold. A small section of the design did not cast properly. Little finishing work is evident other than on the raised border and the center knob. No rotary abrasion is noted on the center knob. Other features include metal ridges from cracks in the mold and a cast pebbled surface. The x-radiograph reveals scattered medium-size porosity throughout the casting.

Condition: The mirror is well preserved with much gray-colored metal evident. On the reflective surface, which has a mottled gray to black appearance, several continuous dots might indicate contact with a plain-weave fabric. The back is essentially a monochromatic gray color; the raised border is slightly darker. Many small, brittle fracture chips are lost from the edge of the reflective side. A blind crack, 5.1 cm long, runs through the middle of the mirror.

方格四乳草葉鏡
Mirror with a Square Band, Four Nipples, and Grass Leaf Motifs
Western Han (206 BC–AD 8)
Weight: 257 g
Diameter: 11.4 cm
Rim height: 0.4 cm
Overall height (knob): 0.9 cm
Inscription: 見日之光，長樂未央 (May you see the light of the sun, and live in everlasting joy.)
1995.284

Manufacture: Cast in high-tin, low-lead bronze in a clay mold. The extent of original finishing work is difficult to determine because of rough cleaning since excavation. Some surface porosity is present.

Condition: The mirror is extensively corroded, with olive green, red, and black corrosion products. The corrosion on the reflective surface was ground down to reveal a mottled surface, exposing some of the red bronze alloy. Scratch marks from the grinding process are present. The back of the mirror was crudely cleaned, leaving many deep scratches and resulting in a mottled black, red/brown, and green surface.

方格四乳草葉鏡 (no. 11)
Mirror with a Square Band, Four Nipples, and Grass Leaf Motifs
Western Han (206 BC–AD 8)
Weight: 809 g
Diameter: 18.3 cm
Rim height: 0.55 cm
Overall height (knob): 1.2 cm
Inscription: 長貴富，宜酒食，樂毋事，日有熹。
1995.285

Manufacture: Cast in high-tin, low-lead bronze in a clay mold. There is some surface porosity; small, evenly distributed porosity is visible in the x-radiograph. Much finishing work is evident. The scalloped border is polished. Rotary abrasion marks are present around the nipples, including the center knob. The background around the inscription was finished, with nearly all the pebbled surface ground away. Some pebbling is seen near the characters. The background is less finished outside the square. The raised leaf shapes are partially chiseled out.

Condition: Much of the silver-colored surface is well preserved. The mirror back has small patches of green and red corrosion, and much burial mud. More corrosion is present on the reflective surface, with approximately 20% of it covered with thick red and green corrosion products and thinner areas of black and gray corrosion. Less burial mud is present on this side. Random post-excavation abrasion marks appear on the reflective surface. A blind crack is present, visible only in the x-radiograph.

雙夔連弧鏡 (no. 35)
Mirror with Linked Arcs and *Kui* Dragons
Eastern Han (25–220)
Weight: 255 g
Diameter: 12.5 cm
Rim height: 0.35 cm
Overall height (knob): 1.3 cm
1995.286

Manufacture: Cast in high-tin, low-lead bronze in a clay mold. Raised areas of the design were abraded and polished; some abrasion marks were not polished out. Rotary finishing marks are present around the central knob. The recessed areas have little to no finishing work, as rough metal from casting can be seen there.

Condition: The reflective surface is covered with green, blue, red, and some black corrosion products that are thick but scattered enough so that some silver-colored metal can be seen. A highly polished surface is preserved in several areas. Hair-like impressions are found in the corrosion layer. More of the silver-colored metal is visible on the back but is not as highly polished. Thick green and red corrosion products as well as a thinner black corrosion product cover approximately one-quarter of the back. Plain-weave textile pseudomorphs and textile fibers reside in corrosion layers near the center knob, and there is burial mud in the recessed areas. A 4.8-cm-long blind crack in the center area of the back has a thin line of green corrosion associated with it.

方格四乳草葉鏡
Mirror with a Square Band, Four Nipples, and Grass Leaf Motifs
Western Han (206 BC–AD 8)
Weight: 143 g
Diameter: 10 cm
Rim height: 0.35 cm
Overall height (knob): 0.75 cm
Inscription: 見日之光，天下大明 (By the light of the sun, the world is made bright.)
1995.287

Manufacture: Cast in high-tin, low-lead bronze in a clay mold. Raised areas of the design were abraded and polished; some abrasion marks are still visible. The recessed areas have little to no finishing, and show rough metal from casting. No finishing work is evident on the center knob.

Condition: The mirror is broken in half with a small loss. Both sides show much gray-colored metal. Thick green corrosion is scattered over the back. The reflective surface is a glossy gray color with scattered thick red and green corrosion products. A streak of vermilion pigment can be seen on the mirror back.

星雲鏡 (no. 13)
Mirror with Clouds and Nebulae
Western Han (206 BC–AD 8)
Weight: 185 g
Diameter: 10 cm
Rim height: 0.4 cm
Overall height (knob): 1.2 cm
1995.288

Manufacture: Cast in high-tin, low-lead bronze in a clay mold. Extensive abrasion and polishing work was done to the raised flat areas. Rotary polishing is evident around the four large nipples and the center knob. Little if any finishing work was done in the recessed areas. Several ridges of metal are present, a result of cracks in the mold. The recessed area on the back is dark gray, while the raised areas are silver-colored.

Condition: Much silver-colored metal is preserved. The back is primarily a silver-bronze color, with scattered areas of red and green corrosion products. Much more of the reflective surface is covered with thick green and red deposits. Thinner black to gray corrosion also is present on the reflective surface. The corrosion products contain an impression of a plain-weave textile.

星雲鏡
Mirror with Clouds and Nebulae
Western Han (206 BC–AD 8)
Weight: 360 g
Diameter: 13 cm
Rim height: 0.6 cm
Overall height (knob): 1.7 cm
1995.289

Manufacture: Cast in high-tin, low-lead bronze in a clay mold. The x-radiograph shows small, evenly scattered porosity throughout. The scalloped edge and raised flat decoration were abraded and polished. The groove near the inner scalloped decoration was partially finished by abrasion. Rotary finishing of four small nipples with a flower-like motif is evident, and extends onto the flat circles nearby. No finishing work is evident on the other nipples.

Condition: Green, red, blue, and black corrosion products almost completely cover the mirror back. Only a few small areas of silver to silver-gray metal can be seen through gaps in the corrosion. Much silver-colored metal is preserved on the reflective surface; approximately one-quarter of the surface is covered with thick layers of green, red, and blue corrosion products. Subsurface corrosion has caused planar distortion of the remaining metal surface. A branched blind crack runs through the middle of the mirror. Small areas of plain-weave textile fibers and pseudomorphs remain on the reflective surface near the edge.

重圈連弧鏡 (no. 14)
Mirror with Concentric Circles and Linked Arcs
Western Han (206 BC–AD 8)
Weight: 132 g
Diameter: 8.1 cm
Rim height: 0.5 cm
Overall height (knob): 0.9 cm
Inscription: 見日之光，天下大明。
1995.290

Manufacture: Cast in high-tin, low-lead bronze in a clay mold. There is some surface porosity. Raised flat areas were polished, while recessed areas were not finished. The center knob was partially finished with rotary abrasion. Metal ridges from mold cracks are only partially removed around the knob.

Condition: Relatively thick green and red as well as thinner black corrosion products cover one-third of the reflective surface, which also shows some random post-excavation abrasion. Relatively thick green and red corrosion products and thin black products cover 10 to 15% of the back. Burial mud is present. Small chip losses have occurred along the edges on both the front and back.

重圈連弧鏡 (no. 15)
Mirror with Concentric Circles and Linked Arcs
Western Han (206 BC–AD 8)
Weight: 421 g
Diameter: 11.9 cm
Rim height: 0.75 cm
Overall height (knob): 1.1 cm
Inscription: 內清以昭明，光象夫日月，不泄。
1995.291

Manufacture: Cast in high-tin, low-lead bronze in a clay mold. Casting defects include mold cracks and chill flow lines; the flow lines appear on the sides of some characters. The border and raised flat areas were smoothed and polished. Some abrasive finishing work was performed on the recessed areas, as fine abrasion marks are visible. Rotary abrasion can be seen around the central knob.

Condition: The mirror has a relatively flat black patina; the back is matte while the reflective side is glossy. There are a number of chips on the edge, especially on the back. The chipped areas also have a black patina. The back is essentially free of corrosion except for a few small patches of green and red deposits. Some light white/gray burial dirt is present. The reflective surface is also essentially corrosion free except for one area of wart-like corrosion with some green corrosion and another small area covered by green corrosion. Red corrosion can also be observed through the breaks in the warts. Deep post-excavation scratches radiate from the area of wart-like corrosion.

重圈連弧鏡
Mirror with Concentric Circles and Linked Arcs
Western Han (206 BC–AD 8)
Weight: 495 g
Diameter: 14.1 cm
Rim height: 0.6 cm
Overall height (knob): 1.3 cm
Inscription: 內清以昭明，光象夫日月，□□ (Its inner purity is shown in perfect reflection. Its light is the image of the sun and moon. . . .)
1995.292

Manufacture: Cast in high-tin, high-lead bronze in a clay mold. Extremely small, evenly distributed porosity is visible in the x-radiograph. Abrasion from finishing work can be observed in gaps in the extensive corrosion on the flat border, on the scalloped band, and on the circles adjacent to the center knob. Little to no finishing work appears to have been done in the low areas of the relief, but thick corrosion layers make determination difficult.

Condition: Thick layers of green, red, and black corrosion cover the mirror. Wart-like subsurface corrosion is evident on both sides. Plain-weave textile pseudomorphs are preserved in the corrosion layers on the mirror back near the edge. Three blind cracks present in the metal can be seen only in the x-radiographs.

鎏金博局式鏡 (no. 25)
TLV Mirror with Gilt Surface
Western Han (206 BC–AD 8)
Weight: 511 g
Diameter: 14.2 cm
Rim height: 0.6 cm
Overall height (knob): 1.2 cm
1995.294

Manufacture: Cast in high-tin, low-lead bronze in a clay mold. The raised band with zigzag decoration was polished. Based on gaps in the gold, little finishing work appears to have been done, as the rough surface would provide better purchase for the gold. The gold is quite thick, suggesting that it was applied as a piece of foil. The nature of the black material is not known. The contrasting colors were an intentional part of the mirror's design.

Condition: The mirror broke roughly in half and was rejoined. Cleaning has exposed much of the silver-gray-colored metal. Most of the corrosion on the reflective surface was abraded off, and many scratch marks are present. An acid or other chemical appears to have been used on the back. There is redeposited pink copper on the gold from previous cleaning procedures. Abrasion and burnishing of the mirror back are also apparent. Small areas of green corrosion remain on both sides, as do many dark gray/black corrosion pits, though the pits are more numerous on the reflective surface. Losses have occurred in the gold foil.

多乳博局式鏡 (no. 27)
TLV Mirror with Multiple Nipples
Xin (9–24)
Weight: 756 g
Diameter: 16.3 cm
Rim height: 0.6 cm
Overall height (knob): 1.2 cm
Inscription: 子丑寅卯辰巳午未申酉戌亥 (inner band); 新有善銅出丹陽，和以銀錫清且明。左龍右虎主四彭，八子 (outer band)
1995.295

Manufacture: Cast in high-tin, low-lead bronze in a clay mold. Finishing work is evident on the back. The T- and V-shapes as well as the central square were chiseled. The center knob and nipples were finished with rotary abrasion, which extends into the four adjoining lobes. The top of the relief decoration was abraded and polished; the recessed areas are largely unfinished, with a pebbled surface and metal ridges caused by mold cracks.

Condition: The back has a glossy black patina. Scattered areas of shallow subsurface green corrosion cover the surface. The reflective surface has a mottled appearance with black and green corrosion products. Areas of black corrosion also have a mottled appearance, containing grays and blacks. Deep scratch marks on the reflective surface indicate that thicker layers of corrosion were abraded after excavation.

博局式鏡
TLV Mirror
Eastern Han (25–220)
Weight: 502.5 g
Diameter: 14.2 cm
Rim height: 0.55 cm
Overall height (knob): 1 cm
Inscription: 來言□□□鏡始，青□在左白虎右。長 (It is said that [the making of this] mirror marks an epoch. At the left is the blue [dragon] and at the right the white tiger. Long . . .)
1995.296

Manufacture: Cast in high-tin, low-lead bronze in a clay mold. The flat raised decoration was abraded and polished. Chisel marks are present in the T- and L-shapes and in the center square. The nipples and center knob were finished with rotary abrasion.

Condition: The mirror was broken into seven pieces, which are reassembled. A number of shallow losses in the reflective side, visible in the x-radiographs, were filled with a less dense material, and a layer of black to brown paint was applied to the reflective surface to hide the joins and fills. On the mirror back, a layer of dirt obscures the joins (some of which do not meet properly). The mirror back has a dense, lustrous black patina layer with small areas of scattered green corrosion. Light, random post-excavation abrasion is evident in the reflective surface's lacquer layer.

四乳四鳥鏡 (no. 18)
Mirror with Four Nipples and Four Birds
Western Han (206 BC–AD 8)
Weight: 93 g
Diameter: 6.2 cm
Rim height: 0.6 cm
Overall height (knob): 0.95 cm
1995.297

Manufacture: Cast in high-tin, low-lead bronze in a clay mold. There is some surface porosity. The concave border band and center knob were abraded and polished. Rotary abrasion was used on the center knob. Little to no finishing work can be seen on other areas of the back.

Condition: Extensive corrosion covers the mirror. The surface is wholly mineralized, and chips on the edge show the interior is also mineralized. A mixture of two subsurface corrosion products give the surface a mottled black and green appearance. Erosion has caused some loss of the reflective surface. Random post-excavation abrasion is visible on the surface of the reflective side.

多乳博局式鏡 (no. 28)
TLV Mirror with Multiple Nipples
Xin (9–24)
Weight: 635 g
Diameter: 16.9 cm
Rim height: 0.5 cm
Overall height (knob): 1.1 cm
Inscription: 子丑寅卯辰巳午未申酉戌亥 (inner band); 作佳鏡哉真大好，上有仙人不知老。渴飲玉泉饑食棗，浮遊天下傲四海。壽如金石為國保 (outer band)
1995.298

Manufacture: Cast in high-tin, low-lead bronze in a ceramic mold. Cracks in the mold can be seen as small ridges of bronze. Cold flow of the metal is also evident. The center square and the T- and L-shapes were finished with chisels. The center knob and nipples, including the base, were polished with rotary action. Background areas between the relief decoration appear to have been carefully scraped without polishing.

Condition: The mirror retains its silver-gray metal color. Only a few small areas of dark gray to black corrosion are present on the back. Streaks of raised green and red corrosion products appear on the reflective surface, along with associated wart-like pits. The reflective side also contains two areas of plain-weave textile pseudomorphs; the textile had very fine, narrow threads. A small crack in the center knob can be seen in the x-radiograph.

多乳博局式鏡 (no. 29)
TLV Mirror with Multiple Nipples
Xin (9–24)
Weight: 549 g
Diameter: 16.6 cm
Rim height: 0.45 cm
Overall height (knob): 1.1 cm
Inscription: 子丑寅卯辰巳午未申酉戌亥 (inner band); 作佳鏡哉真大好，上有仙人不知老。渴飲玉泉饑食棗， 浮遊天下傲四海，壽如金石為國保 (outer band)
1995.299

Manufacture: Cast in high-tin, low-lead bronze in a clay mold. Some finishing work is evident. The design's high areas were abraded and polished. Chisel work is seen in the T- and L-shapes and in the center square. Rotary abrasion can be seen on the nipples. The low areas are largely unfinished.

Condition: The mirror back has a slightly greenish-black lustrous patina; the reflective surface has a mottled black, gray, and green lustrous patina. Fine cracks are present in the reflective surface, the result of subsurface corrosion. Some portions of the reflective surface have flaked off, leaving matte, slightly recessed, areas. Burial mud is trapped in the center knob.

四乳龍虎鏡 (no. 21)
Mirror with Four Nipples, Dragons, and Tigers
Western Han (206 BC–AD 8)
Weight: 572.5 g
Diameter: 4.2 cm
Rim height: 0.7 cm
Overall height (knob): 1.1 cm
1995.300

Manufacture: Cast in high-tin, low-lead bronze in a clay mold. Chill flow lines are clearly visible, especially on the raised linear design. The flat border and midbands were abraded and polished. Rotary abrasion is visible on the four small nipples and around the center knob. Pebbled metal in the recessed areas indicates that no finishing work was done there.

Condition: Much silver-colored metal is preserved. Green, red, and black corrosion products cover approximately three-quarters of the reflective surface; the red and green deposits are quite thick, whereas the black deposits are thin. This surface also contains some wart-like corrosion pits. Approximately 15% of the mirror back—which is darker gray in color than the reflective surface—is covered with green and red corrosion products. Plain-weave textile pseudomorphs are preserved in the corrosion layers, including the thin black layer. The impression of a woven reed mat is seen in the black corrosion layers on the reflective surface. Remains of a circular element (2.5 cm in diameter)—perhaps a coin outline—is found on the reflective side. The metal contains a network of blind cracks. Some red corrosion crystals have grown along the cracks on the back. X-radiographs reveal that the mirror could easily break into four pieces.

四神博局式鏡 (no. 26)
TLV Mirror with Four Spirits and Companions
Western Han (206 BC–AD 8)
Weight: 1,099 g
Diameter: 19 cm
Rim height: 0.7 cm
Overall height (knob): 1.2 cm
Inscription: 子丑寅卯辰巳午未申酉戌亥
1995.301

Manufacture: Cast in high-tin, low-lead bronze in a clay mold. The high areas of the relief decoration were finished by abrasion. A chisel was used to finish the T- and L-shaped channels. The nipples and center knob were polished with rotary abrasion. The four animals on the back are a bright silver color; this difference in color appears to be intentional, as application of silver color over the lines of the animals can be seen under magnification. X-ray fluorescence spectroscopy analyses detect no difference in metal composition between these areas and the rest of the mirror.

Condition: Approximately one-quarter of the reflective surface is silver-colored metal. The remainder is covered with a thick layer of green, blue, red, and black corrosion products as well as thick patches of burial mud. The mirror back has an even black patina with areas of green and red corrosion products.

四葉四乳禽獸鏡
Mirror with Quatrefoil, Four Nipples, Animals, and Birds
Western Han (206 BC–AD 8)
Weight: 733 g
Diameter: 17.4 cm
Rim height: 0.55 cm
Overall height (knob): 1.3 cm
Inscription: 長宜子孫 (May [this mirror] be cherished by your sons and grandsons.)
1995.302

Manufacture: Cast in high-tin, low-lead bronze in a clay mold. Extensive corrosion and subsequent cleaning make determination of finishing work difficult. The flat relief areas were probably finished by abrasion. Rotary abrasion was used on the central knob and four nipples.

Condition: The mirror is extensively corroded. Its back is covered with light green and red corrosion products that were crudely thinned by mechanical cleaning, leaving many scratch marks in the corrosion. Small black granular particles sit atop the corrosion layer. Several blind cracks are present, including one across the central knob. An area on the reflective surface is chipped.

重圈神獸鏡 (no. 22)
Mirror with Concentric Circles, an Immortal, and Auspicious Animals
Western Han (206 BC–AD 8)
Weight: 398 g
Diameter: 13.8 cm
Rim height: 0.5 cm
Overall height (knob): 1 cm
1995.303

Manufacture: Cast in high-tin, low-lead bronze in a clay mold. A ridge of metal resulted from a crack in the mold. Cold flow lines are present in the center area and on the sloped edge leading up to the border. The raised flat areas of the relief decoration were finished by abrasion and polishing. Rotary abrasion was used to finish the nipples and central knob. No chisel marks are evident.

Condition: The mirror has a black to dark gray and green patina. The patina is very glossy on the reflective surface, which contains scattered wart-like corrosion pits. The mirror back is more matte and also has scattered wart-like pits.

重圈神獸鏡 (no. 23)
Mirror with Concentric Circles, an Immortal, and Auspicious Animals
Eastern Han (25–220)
Weight: 719 g
Diameter: 16.3 cm
Rim height: 0.95 cm
Overall height (knob): 1.5 cm
Inscription: 宜子
1995.304

Manufacture: Cast in high-tin, low-lead bronze in a clay mold. The flat raised relief decoration was highly finished by abrasion and polishing; the nipples were finished with rotary grinding. Little finishing was done in the background area.

Condition: A network of large blind cracks exists in the middle of the mirror, with one extending to the rim; some small surface losses are associated with this crack. The mirror is extensively mineralized. Some burial mud and a thin gray-black patina with thick patches of green corrosion are found on the back; green corrosion on the central knob preserves plain-weave textile pseudomorphs. The reflective surface has a thicker glossy black corrosion product with some gray-colored mottling, as well as scattered green corrosion pits.

四乳八禽鏡 (no. 19)
Mirror with Four Nipples and Eight Birds
Western Han (206 BC–AD 8)
Weight: 236 g
Diameter: 9.6 cm
Rim height: 0.65 cm
Overall height (knob): 1 cm
1995.305

Manufacture: Cast in high-tin, low-lead bronze in a clay mold. The flat border, four small nipples, and central knob have all been finished by abrasion. Rotary action was used on the nipples and knob. No finishing is evident in the background.

Condition: The overall glossy black patina is relatively even, with some gray mottling on the reflective surface. This surface, which contains three deep post-burial scratches, has a higher gloss than the back. A fine network of cracks—a result of the corrosion process—is evident in the reflective surface. The corrosion is brittle, resulting in several chips at the surface edge. Several smaller chips are present on the edge of the back.

重圈連弧鏡 (no. 16)
Mirror with Concentric Circles and Linked Arcs
Western Han (206 BC–AD 8)
Weight: 566 g
Diameter: 14.9 cm
Rim height: 0.7 cm
Overall height (knob): 1.1 cm
Inscription: 湅治銅華清而明，以之為鏡而宜文章。延年益壽而辟不羊。與天毋極。
1995.306

Manufacture: Cast in high-tin, low-lead bronze in a ceramic mold. Cracks in the mold caused small ridges to form in the metal. The back was carefully polished, leaving no tool marks or abrasion marks. The knob was polished with rotary motion, especially at the base.

Condition: The mirror has a mottled black and green patina. All corrosion is subsurface. Light, random post-burial abrasion is visible on the reflective side.

重圈連弧鏡
Mirror with Concentric Circles and Linked Arcs
Western Han (206 BC–AD 8)
Weight: 60.5 g
Diameter: 6.9 cm
Rim height: 0.45 cm
Overall height (knob): 0.7 cm
Inscription: 見日之光，天下大明 (By the light of the sun, the world is made bright.)
1995.307

Manufacture: Cast in high-tin bronze in a clay mold. Some medium-size porosity is scattered throughout the casting. The raised flat areas of the design were polished, but little finishing work can be seen in the recessed areas. Rotary polishing around the center knob is evident.

Condition: The mirror has a relatively smooth, glossy black patina. Part of the reflective surface exhibits pitting corrosion and some wart-like corrosion; post-burial scratches also are evident. Both sides of the mirror have small scattered areas of green corrosion.

四乳四虺鳥紋鏡
Mirror with Four Nipples, Quasi-Dragons, and Birds
Western Han (206 BC–AD 8)
Weight: 93 g
Diameter: 8.8 cm
Rim height: 0.3 cm
Overall height (knob): 0.85 cm
1995.308

Manufacture: Cast in high-tin, low-lead bronze in a clay mold. Some surface porosity is observed on the back edge band, center knob, and reflective surface. Some abrasive finishing work is evident on the flat outer border. The four nipples and center knob were finished with rotary abrasion. There is no other evidence of finishing.

Condition: The mirror has a green-black patina, which is even on the back and more mottled on the reflective surface. The corrosion products are hard, compact, and brittle. There are several brittle chips and losses from the reflective side. Under magnification, a fine network of surface cracks from the corrosion process can be seen on both sides of the mirror.

重圈連弧鏡
Mirror with Concentric Circles and Linked Arcs
Western Han (206 BC–AD 8)
Weight: 37 g
Diameter: 6.2 cm
Rim height: 0.3 cm
Overall height (knob): 0.7 cm
Inscription: 見日之光，天下大明 (By the light of the sun, the world is made bright.)
1995.309

Manufacture: Cast in high-tin, low-lead bronze in a clay mold. Some porosity exists on the raised border on the back and on the reflective surface. The raised flat areas of the design were abraded and polished. Some rotary abrasion is present near the center knob. Little to no finishing work is evident in the recessed areas.

Condition: A gray-colored metal surface is preserved. Scattered green, red, and black corrosion products are on the back, along with some burial dirt trapped in recessed areas. The reflective surface contains scattered black and gray corrosion, green corrosion below the surface in pits, and some deep post-excavation scratches. A blind crack extends 1 cm in from the edge and takes a 90-degree turn for another 1.6 cm.

重圈紋鏡
Mirror with Concentric Circles
Western Han (206 BC–AD 8)
Weight: 34 g
Diameter: 5.9 cm
Rim height: 0.3 cm
Overall height (knob): 0.6 cm
Inscription: 見日之光，天下大明 (By the light of the sun, the world is made bright.)
1995.310

Manufacture: Cast in high-tin, low-lead bronze in a clay mold. Fine, evenly distributed porosity is visible in the x-radiograph. Raised areas of the relief were finished, with abrasion marks found in several areas. The degree of finishing work in the recessed areas is difficult to determine due to extensive corrosion.

Condition: Thick green, red, and black corrosion products cover more than 95% of the back surface. One area exhibits a white accretion, with accompanying scratch marks from attempted cleaning. Approximately three-quarters of the reflective surface is covered with thick deposits of red, green, and black corrosion products, along with some wart-like corrosion and pits.

四乳四虺鳥紋鏡
Mirror with Four Nipples, Quasi-Dragons, and Birds
Western Han (206 BC–AD 8)
Weight: 544 g
Diameter: 14.6 cm
Rim height: 0.6 cm
Overall height (knob): 1.4 cm
1995.311

Manufacture: Cast in high-tin, low-lead bronze in a clay mold. Little finishing work is evident other than on raised flat areas that were abraded and polished. The four nipples and central knob were finished with rotary abrasion. Cracks in the clay mold caused small ridges of metal on the back.

Condition: The mirror shows areas of bright silver-colored metal. Approximately one-quarter of the reflective surface and one-third of the back are a silver color. The remainder of the object is covered with thick red and green corrosion products and a thinner black corrosion product. Several small areas with a plain-weave textile pseudomorph and perhaps some fibers are preserved at the edge. The x-radiograph shows a blind crack bisecting the mirror.

重圈紋鏡
Mirror with Concentric Circles
Western Han (206 BC–AD 8)
Weight: 89 g
Diameter: 7.7 cm
Rim height: 0.4 cm
Overall height (knob): 0.7 cm
Inscription: 見日之光，長毋相忘 (By the light of the sun, forget me not—ever.)
1995.312

Manufacture: Cast in high-tin, low-lead bronze in a clay mold. Some abrasion and polishing are evident on the flat edge and midband. Rotary abrasion is present around the central knob. No finishing is visible in the background and relief decoration.

Condition: The reflective surface has extensive thick green, red, and blue corrosion layers, in which are preserved textile fibers and plain-weave pseudomorphs. Approximately half of the back preserves silver-colored metal. Thin black and gray along with thick green and red corrosion products are scattered over the remaining surface. A 2.2-cm-long blind crack extends from the open end of the knob adjacent to the silver metal to the middle of the inscription circle.

四乳四虺鳥紋鏡
Mirror with Four Nipples, Quasi-Dragons, and Birds
Western Han (206 BC–AD 8)
Weight: 95 g
Diameter: 8.9 cm
Rim height: 0.35 cm
Overall height (knob): 0.8 cm
1995.313

Manufacture: Cast in high-tin, low-lead bronze in a ceramic mold. The raised border was smoothed and polished. Rotary polishing marks can be seen around the four nipples and the central knob. There is no evidence of finishing work in the recessed areas.

Condition: Hard, brittle, compact, mottled black and green corrosion products cover the surface of the mirror, which is entirely mineralized. A chip in the edge of the mirror was caused by the brittle nature of the corroded surface. The x-radiograph reveals much internal corrosion.

四乳四虺鳥紋鏡 (no. 20)
Mirror with Four Nipples, Quasi-Dragons, and Birds
Western Han (206 BC–AD 8)
Weight: 173 g
Diameter: 9.9 cm
Rim height: 0.4 cm
Overall height (knob): 0.9 cm
1995.314

Manufacture: Cast in high-tin, low-lead bronze in a clay mold. X-radiographs reveal little porosity and, except in the center knob, the pores have a very small diameter. Finishing work is evident on the back. The flat bands were polished and the central knob and nipples were polished with rotary action. The flat design areas nearly match the reflective surface in degree of polish. Little finishing work can be seen in the low areas around the relief decoration.

Condition: The mirror has a glossy black appearance. The back is mottled with thin surface red and green corrosion products and burial dirt. The metal on the rim has some wart-like subsurface corrosion. The reflective surface, which contains some mild post-burial scratches, is mottled with gray, black, green, and red surface corrosion products. A pebbled textile pattern is visible in the black and gray corrosion; the regularity and equal spacing of the peaks indicate the textile had a plain weave.

雙鳳鏡
Mirror with Paired Phoenixes
Eastern Han (25–220) to Western Jin (265–316)
Weight: 121 g
Diameter: 9.8 cm
Rim height: 0.4 cm
Overall height (knob): 1 cm
Inscription: 位至三公 (May you attain dukedom!)
1995.315

Manufacture: Cast in high-tin, low-lead bronze in a clay mold. A Y-shaped ridge of metal resulting from a crack in the mold bisects the back. The flat border was scraped and polished. Little to no finishing work was done on the background or the relief decoration. The central knob was not finished. The knob has a pebbled surface, perhaps an indication that the metal may have solidified prematurely.

Condition: Much of the reflective surface is covered with thick layers of green and red corrosion products interspersed with silver metal mottled with black and gray corrosion. The back has green, black, blue, and red corrosion products. Some gray-colored metal is observed on the border; some red burial dirt clings to the surface. Random post-excavation scratches are seen on the edge band. The corrosion layers on each side of the mirror preserve plain-weave textile pseudomorphs.

雙鳳鏡
Mirror with Paired Phoenixes
Eastern Han (25–220) to Western Jin (265–316)
Weight: 58 g
Diameter: 7.7 cm
Rim height: 0.15 cm
Overall height (knob): 0.8 cm
Inscription: 位至三公 (May you attain dukedom!)
1995.316

Manufacture: Cast in high-tin bronze in a clay mold. Some medium-size porosity is visible in the x-radiograph. The border was extensively finished by scraping and abrasion. Little finishing work was carried out in the design area.

Condition: Much uncorroded silver-colored metal is visible on the mirror, with a few areas of green, blue, and red corrosion products scattered on both the front and back. Some burial mud resides in the design's recessed areas, and post-excavation scratches are visible throughout. Under magnification the cast dendritic structure can be seen in the raised decoration. X-radiographs reveal two blind cracks on either side of the knob.

四葉瑞獸對鳳鏡 (no. 37)
Mirror with Quatrefoil, Paired Phoenixes, and Auspicious Animals
Three Kingdoms (220–280) to Western Jin (265–316)
Weight: 694 g
Diameter: 18.6 cm
Rim height: 0.4 cm
Overall height (knob): 1.2 cm
1995.317

Manufacture: Cast in high-tin, low-lead bronze in a clay mold. Some porosity is visible in the x-radiograph, and both sides of the mirror exhibit much surface porosity. Abrasion marks on some raised parts of the decoration indicate that the relief decoration was finished. No finishing work is evident in the recessed areas.

Condition: The mirror is extensively corroded, its back covered with a thick olive green corrosion product. The corrosion process exposed the cast dendritic structure. The reflective side, which is mottled with smooth subsurface green, gray, and black corrosion products, retains a glossy reflective surface with some random post-excavation scratches. X-radiographs reveal a network of blind cracks surrounding the central knob.

四葉重圈連弧紋鏡 (no. 17)
Mirror with Quatrefoil, Concentric Circles, and Linked Arcs
Eastern Han (25–220)
Weight: 406 g
Diameter: 15 cm
Rim height: 0.5 cm
Overall height (knob): 1.2 cm
Inscription: 長宜子孫
1995.318

Manufacture: Cast in high-tin, low-lead bronze in a clay mold. Scattered porosity is seen in the x-radiograph. The raised areas of the decoration were abraded and polished; the metal in the recessed areas is not as well finished. There is evidence of rotary grinding at the center knob's base.

Condition: For the most part, the mirror remains a bright silver-gray color. Areas of raised red and green corrosion are present on both sides. The heaviest corrosion on the back is around the perimeter. Several small areas of black corrosion exist on the reflective surface. On both sides, plain-weave textile pseudomorphs are present in some of the green corrosion.

四葉連弧鏡 (no. 36)
Mirror with Quatrefoil and Linked Arcs
Eastern Han (25–220)
Weight: 235 g
Diameter: 12.2 cm
Rim height: 0.3 cm
Overall height (knob): 1.1 cm
Inscription: 君宜官位
1995.319

Manufacture: Cast in high-tin, low-lead bronze in a clay mold. The raised areas of relief decoration were ground and polished. The metal in the recessed areas is rough, indicating little finishing work there. Uneven metal flow or mold damage is visible in the channel near the border.

Condition: Some of the surface is preserved as a dark gray-colored metal. Approximately 35% of the back and 10% of the reflective surface are a dark silver color. Thick green and red corrosion, with some thin black/gray-colored corrosion layers, is present on both sides. The corrosion products on the reflective surface were thinned or reduced mechanically; the corrosion was probably more wart-like before the removal attempts. In the back near the knob is a blind crack, approximately 3.7 cm long. Red paint found on the knob top may be a Chinese character.

雙夔鏡 (no. 34)
Mirror with Linked Arcs and *Kui* Dragons
Eastern Han (25–220)
Weight: 465 g
Diameter: 15.3 cm
Rim height: 0.45 cm
Overall height (knob): 1.2 cm
Inscription: 潘氏作文兮
1995.320

Manufacture: Cast in high-tin, low-lead bronze in a clay mold. Extensive finishing work was done on the back. Rotary abrasion is visible at the base of the knob. The raised areas were smoothed and polished, with less finishing work in the recessed areas. One dragon head contains a ridge of metal from a crack in the mold.

Condition: The mirror back has an even, glossy black patina with very few scattered areas of green corrosion. A mottled green and black surface characterizes the reflective side. A branched T-shaped crack extends from one edge into the center region and exhibits some chip losses along its length.

瑞獸鏡
Mirror with Four *Qilin*
Eastern Han (25–220)
Weight: 254.5 g
Diameter: 10.7 cm
Rim height: 0.7 cm
Overall height (knob): 1.1 cm.
Inscription: 吾作明鏡世少有，服之吉利宜孫子。樂兮。
1995.321

Manufacture: Cast in high-tin, low-lead bronze in a ceramic mold. The raised flat areas of the design were ground and polished. The nipples and central knob were polished by rotary abrasion.

Condition: The mirror's reflective surface is primarily mineralized with some gray metallic areas. The corrosion is thick and often wart-like, penetrating deep beneath the surface. Both sides have wart-like pits with interior red corrosion surrounded by green corrosion. A crack at the edge penetrates completely through the metal. There is a network of shallower cracks throughout the mirror.

龍虎對峙鏡 (no. 42)
Mirror with Two Dragons and a Tiger
Three Kingdoms (220–280)
Weight: 770 g
Diameter: 14.7 cm
Rim height: 1.1 cm
Overall height (knob): 1.5 cm
Inscription: 張氏作鏡四夷服，多賀君家人民息。胡羌殄滅天下復，風雨時節五穀熟。官至公卿蒙祿食，長保二親子孫力。樂毋極
1995.322

Manufacture: Cast in high-tin, low-lead bronze in a clay mold. A mold crack appears as a metal ridge along the step between the center area and the border. The x-radiograph reveals some porosity at the edge. The raised flat decoration was smoothed and polished. Evidence of rotary abrasion is seen on the center knob. Less finishing work was performed in the recessed areas.

Condition: The mirror back has a relatively even black patina with scattered areas of green and red corrosion. There is an area on the step between the center and the raised rim (adjacent to one of the dragons) where extensive corrosion has caused the metal to expand, distorting the surface. The reflective surface has a glossy black patina mottled with some gray. Green corrosion has a wiped appearance but is actually below the surface. One or two areas of wart-like corrosion are present near the edge. The corrosion process has caused fine cracks in the reflective surface. Internal cracks are visible in normal light and in the x-radiograph.

盤龍鏡 (no. 43)
Mirror with a Coiling Dragon
Three Kingdoms (220–280)
Weight: 222 g
Diameter: 9.9 cm
Rim height: 0.8 cm
Overall height (knob): 1 cm
1995.323

Manufacture: Cast in high-tin, low-lead bronze in a clay mold. Much porosity can be seen at the mirror edges. Cracks from the mold are apparent in the metal and cold flow lines visible over the surface. Moderate finishing work with abrasion and polishing of raised areas was performed on the mirror back. The knob was finished with rotary grinding and polishing. The recessed areas are not finished.

Condition: The mirror is well preserved with a gray to black-colored patina. The patina on the reflective surface is markedly mottled, with small areas of green corrosion and much fine random post-excavation abrasion.

半圓方枚環繞式瑞獸鏡 (no. 40)
Mirror with Auspicious Animals Surrounded by Rings of Squares and Semicircles
Eastern Han (25–220)
Weight: 242 g
Diameter: 11.8 cm
Rim height: 0.4 cm
Overall height (knob): 0.8 cm
Inscription: 吾作明鏡，幽湅三商。伯牙作樂，眾神見容。宜大福兮，有福安服。曾年宜壽，子孫番昌。其師命長。
1995.324

Manufacture: Cast in high-tin, low-lead bronze in a clay mold. An area of porosity is visible at one edge. An adjacent area was repaired by inserting a piece of metal. Several cracks in the mold appear on the mirror back as metal ridges. The raised flat areas of the decoration were carefully smoothed and polished. Less finishing work was done in the recessed areas. Rotary grinding is evident on the center knob.

Condition: The mirror back has a black to greenish-black patina, with areas of green corrosion around the edges. The reflective surface is a mottled black, green, and light green. Evidence of subsurface corrosion appears as concave scales on the metal surface. Areas of lost design were built up and reinforced by applying a pigmented resin. A number of the raised dots were added and several lines on the dragon were reinforced. An internal blind crack forms a V-shape around the center knob.

半圓方枚環繞式神獸鏡
Mirror with Deities and Animals Surrounded by Rings of Squares and Semicircles
Eastern Han (25–220)?
Weight: 292 g
Diameter: 10.8 cm
Rim height: 0.5 cm
Overall height (knob): 1.05 cm
Inscription: [characters indecipherable]
1995.325

Manufacture: Cast in low-tin, high-lead bronze by the lost-wax method (?). Extensive corrosion makes it difficult to determine the casting process and the extent of finishing work. Some porosity is observed in the x-radiograph. The poor quality of the cast details and the high-lead alloy indicates that this mirror was recast.

Condition: Extensive thick green and red corrosion layers cover the back, on which a glossy coating was applied. The reflective surface also is covered with thick layers of green, red, and black corrosion products. A wax-like brown to black coating was applied to the reflective surface, obscuring much of the corrosion. Some small points of silver-colored metal are scattered over the reflective surface. Two blind cracks in the mirror can be seen only in the x-radiograph.

半圓方枚環繞式神獸鏡 (no. 39)
Mirror with Deities and Animals Surrounded by Rings of Squares and Semicircles
Eastern Han (25–220)
Weight: 270 g
Diameter: 11.8 cm
Rim height: 0.35 cm
Overall height (knob): 1.2 cm
Inscription: 吾作明鏡，幽湅三商。雕刻極無。
1995.326

Manufacture: Cast in high-tin, low-lead bronze in a clay mold. Some finishing work is visible on the shallow relief border, along with some abrasion marks. Rotary abrasion appears around the center knob. No other finishing work is evident.

Condition: The mirror has a thin black patina on the back and a mottled gray and black patina on the reflective surface. Some small light green pits are present on the reflective surface, along with a network of fine surface cracks. Areas of green and red corrosion appear on the back. The x-radiograph reveals two blind cracks—one in the center area and one at the edge. Wax or a similar material was applied to the mirror's surface.

瑞獸鏡 (no. 45)
Mirror with Four *Pixie*
Northern and Southern Dynasties (420–589)
Weight: 347 g
Diameter: 11.6 cm
Rim height: 0.6 cm
Overall height (knob): 1.3 cm
Inscription: 吾作明鏡，幽湅三商。雕刻典祚，配德萬疆。人氣顯分，天元景容。座事六采，光耀玄黃，統德序道，□有紀綱...
1995.327

Manufacture: Cast in high-tin, low-lead bronze in a clay mold. Limited finishing work is evident. The flat relief decoration at the edge was ground down. Rotary abrasion is evident at the center knob and at the triangular points.

Condition: Extensive corrosion has occurred. The reflective surface is mottled light green and black. The light green corrosion is thin and brittle; the black corrosion is glossy except where the surface layer is lost in an area near the edge. Wart-like corrosion is present on the reflective surface. The back is primarily a black color with scattered green surface corrosion. Extensive pitting and loss of metal is visible near the edge. X-radiographs reveal a blind crack adjacent to the knob.

半圓方枚環繞式三段神人鏡 (no. 41)
Mirror Featuring Deities and Kings in Three Sections Surrounded by Rings of Squares and Semicircles
Eastern Han (25–220)
Weight: 280.5 g
Diameter: 13.9 cm
Rim height: 0.4 cm
Overall height (knob): 1 cm
Inscription: 吾作明鏡，幽湅三岡。巧工刻之。
1995.328

Manufacture: Cast in high-tin, low-lead bronze in a clay mold. The head of the Queen Mother of the West did not cast. On the back, a small ridge of metal caused by a mold crack is visible in the upper left and lower right quadrants. Little finishing work can be found except on the raised relief decoration at the edge, with some rotary abrasion around the central knob.

Condition: Much uncorroded silver-colored metal is preserved, including approximately three-quarters of the back and one-third of the reflective surface. The back has areas of green, black/gray, and red corrosion products. The reflective surface is covered with thick green and red corrosion with thinner black corrosion products, and shows extensive deep post-excavation abrasion of the corrosion. This deep abrasion extends slightly into the silver-colored metal, which also exhibits fainter random abrasion. A faint area of plain weave textile pseudomorphs is preserved near the edge.

方格神獸鏡 (no. 30)
Mirror with a Central Square, an Immortal, and Auspicious Animals
Eastern Han (25–220)
Weight: 620 g
Diameter: 18.5 cm
Rim height: 0.8 cm
Overall height (knob): 1.7 cm
1995.329

Manufacture: Cast in high-tin, low-lead bronze in a ceramic mold. A moderate amount of finishing work was performed. The design's raised flat areas were polished. There is evidence of rotary abrasive work around the four nipples and at the base of the central knob. No chisel work is apparent in either of the concentric channels or the square. The recessed areas show little to no finishing work.

Condition: The mirror back is gray in color, mottled with some black around the perimeter. Several areas of green corrosion are scattered over the surface. The reflective surface has a mottled appearance, ranging from a dark green-black to a pale green color. Areas of red corrosion are present in the central area, some rust stains can be seen on one side. Several scattered deep pits, and a network of cracks from the corrosion process span the surface. Two areas suggest that the mirror was in contact with a textile. The corrosion patterns on the knob top and one area of red corrosion on the back suggest plain-weave textiles. The x-radiograph reveals a star-shaped crack near the center knob, which is nearly encircled by one leg of the crack. A second crack is near one of the nipples.

方格神獸鏡 (no. 31)
Mirror with a Central Square, Immortals, and Auspicious Animals
Eastern Han (25–220)
Weight: 643 g
Diameter: 18.35 cm
Rim height: 0.7 cm
Overall height (knob): 2 cm
1995.330

Manufacture: Cast in high-tin, low-lead bronze in a clay mold. The degree of finishing work is difficult to determine because of overpaint. The flat relief decoration at the edge was ground and polished. Rotary abrasion was used to finish the nipples and center knob.

Condition: The x-radiograph reveals that the mirror was broken into at least four pieces, with an extensive network of cracks at the center. Some cracks or breaks are visible in the metal on the mirror back. The breaks are hidden by extensive black overpaint. Black spray paint can be observed on the dirt trapped in the hole of the center knob. Some light green-colored corrosion was left visible on the mirror back. The light green substance on the reflective side is applied paint.

畫像鏡 (no. 32)
Mirror with Deities and Kings
Eastern Han (25–220)
Weight: 847 g
Diameter: 19.5 cm
Rim height: 1 cm
Overall height (knob): 1.9 cm
Inscription: 王母。玉女。
1995.331

Manufacture: Cast in high-tin, low-lead bronze in a ceramic mold. The x-radiograph reveals some porosity in the metal. Because of the object's condition and the thick coating that was applied, it is difficult to determine the degree of finishing done to the mirror back. It is clear that the nipples and knobs were polished by rotary action.

Condition: The x-radiograph reveals that the mirror was broken into six pieces, with five large losses along the edge. These losses were filled with a material less radio-opaque than the bronze, and an opaque black coating was applied to hide the damage; the black also covers burial dirt in the hole in the center knob. The coating is less opaque on the reflective side than on the back, and magnification shows some areas of corrosion.

畫像鏡 (no. 33)
Mirror with Deities, Chariot, and the White Tiger
Eastern Han (25–220)
Weight: 503 g
Diameter: 18.3 cm
Rim height: 0.85 cm
Overall height (knob): 1.7 cm
Inscription: 東王公。白虎。
1995.332

Manufacture: Cast in high-tin, low-lead bronze in a ceramic mold. The extent of finishing work is difficult to determine because of the object's condition.

Condition: The mirror was shattered into more than two dozen pieces and rejoined. X-radiographs reveal four areas of loss filled with a material less dense than bronze. Paint was applied to disguise the damage. Under magnification some areas of the decoration appear to be built up in a light brown-colored resin. Both sides of the mirror display extensive abrasion. The reflective surface was ground and polished after excavation. The mirror back has deep scratches and abrasion, especially around the center knob. This abrasion is probably the result of restoration and removal of a thick corrosion layer.

半圓方枚環繞式神獸鏡 (no. 38)
Mirror with Deities and Animals Surrounded by Rings of Squares and Semicircles
Eastern Han (25–220)
Weight: 509 g
Diameter: 15 cm
Rim height: 0.45 cm
Overall height (knob): 1.3 cm
Inscription: 吾作明鏡，幽涑三岡。配仙世袁，
統德序道。敬奉臣良，同刻無極。百牙作樂，
眾祥主陽。世照光明，壽如金石。位至三公，
其師命長。
1995.333

Manufacture: Cast in high-tin bronze in a ceramic mold. Abrasive finishing work is evident on the raised flat areas and in the recessed wells. The wells may have been finished with rotary abrasion, but determination is difficult due to post-excavation surface scratches. Small ridges of metal resulted from cracks in the mold. Little finishing work is evident in the background.

Condition: The bronze surface is almost entirely mineralized. The reflective surface is glossy mottled gray to black, scattered with lifting blisters and pits from the corrosion process. A Y-shaped crack in the metal has an associated network of shallow cracks. The mirror back appears to have a thin colored coating which makes this side more monochromatic. Several small areas on the back have been reconstructed using a pigmented resin.

四乳八獸鏡 (no. 44)
Mirror with Four Nipples and Eight Animals
Western Jin (265–316), dated AD 300
Weight: 1,495 g
Diameter: 19.6 cm
Rim height: 1 cm
Overall height (knob): 1.3 cm
Inscription: 永康元年正月午日，由[涑]黃日。
早作明鏡，大吉大宮。延長壽命，上有王父西王母。
1995.334

Manufacture: Cast in high-tin, low-lead (2 to 3% zinc) bronze in a clay mold. The extent of finishing work is difficult to determine because of a thick layer of black paint and evidence of post-burial abrasion (wire brushing) on the decorative surface. It is likely that the flat raised edge decoration was carefully finished, and the nipples were probably finished with rotary abrasion.

Condition: Both sides have a thick layer of black paint. X-radiographs reveal that the mirror was broken into seven pieces that were rejoined with lead-based solder. Small triangular fills were made at the ends of the breaks on the mirror edge. Some green corrosion is evident on the mirror back in the more protected details. The back was abraded with a wire-wheel brush, probably to reduce the corrosion. Four deep, pronounced scratches appear in the paint (lacquer) on the reflective surface. Numerous fine random scratches are also present in the paint on the reflective surface.

瑞獸鏡 (no. 51)
Mirror with Three Pairs of *Pixie* around a Loti-Form Knob
Tang (618–907)
Weight: 989 g
Diameter: 17.6 cm
Rim height: 1.3 cm
Overall height (knob): 1.5 cm
Inscription: 鍊形神冶，瑩質良工。如珠出匣，
似月停空。當眉寫翠，對臉傳紅。 綺窗繡幌，
俱含影中。
1995.335

Manufacture: Cast in high-tin bronze by the lost-wax method. The back is carefully finished, with polished relief and background areas.

Condition: Much of the mirror surface is metallic silver/gray in color. Random areas of green corrosion appear on the mirror back, while much of the polished surface is covered with thick patches of green and red corrosion products. Six major cracks exist, along with a network of shallower cracks throughout. Corrosion products are visible inside the cracks.

纏枝花十二生肖鏡 (no. 46)
Mirror with Interlacing Floral Motifs and Twelve Zodiac Animals
Sui (581–618)
Weight: 294 g
Diameter: 15.4 cm
Rim height: 0.45 cm
Overall height (knob): 1.4 cm
Inscription: 光正隨人，長命宜新。
1995.336

Manufacture: Cast in high-tin, low-lead bronze by the lost-wax method. The design is fluid and appears to be built up in wax. Little to no finishing work was done to the mirror back.

Condition: Some uncorroded silver-colored metal can be observed, but the majority of the surface is covered with green, black, gray, and red corrosion products. Some blue corrosion and wart-like corrosion are on the reflective surface. The ridges of the relief decoration are black, while the recessed areas are more silver. X-ray fluorescence analysis reveals the presence of mercury on both surfaces.

重圈神獸鏡 (no. 47)
Mirror with Deities and Animals in Concentric Circles
Sui (581–618)
Weight: 1,393 g
Diameter: 33.4 cm
Rim height: 0.9 cm
Overall height (knob): 1.6 cm
Inscription: 光正隨人，長命宜新。
1995.337

Manufacture: Cast in high-tin, low-lead bronze by the lost-wax method. Fine, evenly distributed porosity can be seen in the x-radiographs. There is little finishing work on the mirror back except to the highest relief areas. No finishing work was done in the recessed wells of the design, which are filled in with rough, irregular metal.

Condition: The mirror is predominantly a gray-silver metal color. Patches of green and red corrosion are scattered over the back. On the reflective surface the corrosion is more dense and includes some small wart-like pits. Plain-weave textile pseudomorphs are preserved in the corrosion layers, and burial dirt is present in recessed areas.

瑞獸鏡 (no. 50)
Mirror with Four Running Animals
Tang (618–907)
Weight: 156 g
Diameter: 9.6 cm
Rim height: 0.7 cm
Overall height (knob): 0.9 cm
Inscription: 賞得秦王鏡，判不惜千金。非關欲照膽，特是自明心。
1995.338

Manufacture: Cast in high-tin, low-lead bronze by the lost-wax method. The high area of the relief decoration was finished by abrasion. Little evidence of finishing work is seen in the background. The knob is polished, but there is no trace of rotary abrasion.

Condition: The mirror back—metallic silver in color—is virtually corrosion free. The reflective surface, approximately half of which is metallic silver-colored metal, has patches of thick green, red, and black corrosion products as well as areas of wart-like corrosion.

四神鏡 (no. 49)
Mirror with Four Spirits
Sui (581–618)
Weight: 870 g
Diameter: 19.8 cm
Rim height: 0.65 cm
Overall height (knob): 1.5 cm
Inscription: 仙山並照，智水齊名。花朝艷采，月夜流明。龍盤五瑞，鸞舞雙情。傳聞仁壽，始驗銷兵。
1995.339

Manufacture: Cast in high-tin, low-lead bronze by the lost-wax method. Minimal finishing was done on the back surface. Much spongy surface structure from casting is visible. The high areas of the relief decoration were polished.

Condition: This well-preserved mirror has much silver-colored metal. Areas of green, red, and black corrosion products are on the back. The reflective surface has a mottled black and gray-colored patina with scattered areas of green, red, and blue corrosion as well as random post-burial scratches.

瑞獸葡萄鏡 (no. 52)
Animal-and-Grape Mirror
Tang (618–907)
Weight: 576.5 g
Diameter: 11.7 cm
Rim height: 1.2 cm
Overall height (knob): 1.3 cm
1995.340

Manufacture: Cast in high-tin, low-lead bronze by the lost-wax method. Extensive abrasive finishing is evident in the background and in the relief decoration. Cast features were somewhat softened by finishing. Spacing of the triangular points on the vertical wall was adjusted to fit the available area.

Condition: The metal contains two cracks which are more easily seen in the x-radiograph, since they were filled and inpainted with aluminum-based pigment or paint (visible in several places). One triangular chip loss at the edge was also filled and inpainted. The reflective side is covered with thick layers of red and green corrosion products, as well as several areas of thinner black and gray corrosion. Some grinding of the corrosion products has abraded the surface. Green, red, and black corrosion products partially cover half of the back. Much silver-colored metal is visible in corrosion-free areas.

寶相花鏡 (no. 57)
Mirror with Six Circular Flowers
Tang (618–907)
Weight: 833 g
Diameter: 16.6 cm
Rim height: 0.9 cm
Overall height (knob): 1.5 cm
Inscription: □□□膽，知幽察微。珠漸朗潤，月謝光輝。□兵既弭，福慶斯歸。
1995.341

Manufacture: Cast in high-tin, low-lead bronze by the lost-wax method. The back was carefully finished by abrasion and polishing of the background and high relief areas. Chisel marks are not found.

Condition: Much of the mirror's silver metal color is preserved. Green, red, and blue (seen under magnification) corrosion products cover less than one-quarter of the back. The reflective surface is more heavily corroded; half of it is covered with thick layers of green and red corrosion. Patches of black to gray-black corrosion appear on the more metallic areas. Large wart-like corrosion pits indicate substantial subsurface corrosion. Broken pits reveal a red-colored corrosion product inside. Textile pseudomorphs as well as actual textile fibers—both plain weave and gauze—are preserved in the green corrosion layers. The gauze is on the reflective surface. The top of the center knob is cracked, perhaps from expansion of the metal during the corrosion process. A faint surface crack is visible on the reflective side.

雙禽瑞獸花枝鏡
Octafoil Mirror with Crouched Animals
Tang (618–907)
Weight: 1,074 g
Diameter: 19.3 cm
Rim height: 0.95 cm
Knob height: 0.9 cm
1995.342

Manufacture: Cast in high-tin, low-lead bronze by the lost-wax method. Extensive finishing of the background is evident, including chisel marks and polishing. Little finishing work was performed on the relief decoration.

Condition: Most of the corrosion products were removed, perhaps by electrolytic reduction, leaving a pink redeposited copper in the corroded areas. Approximately half the back is copper-colored, while the remainder is silver-colored. The reflective surface is copper-colored with mottled areas of green and red corrosion. A number of deep gouges in the metal resulted from the corrosion removal process. Cleaning produced visible scratches on the mirror back.

雙童戲荷葵花鏡
Lobed Mirror with Acrobats on Lotus Blossoms
Tang (618–907)
Weight: 527 g
Diameter: 14.1 cm
Rim height: 0.6 cm
Overall height (knob): 0.8 cm
1995.343

Manufacture: Cast in high-tin, low-lead bronze by the lost-wax method. Extensive finishing of the back is evident. The background was carefully finished with chisel work and polishing up to the relief decoration. The high areas of relief were polished, softening the details. A rough-cast metal surface is visible inside the center knob.

Condition: The mirror is extensively corroded. Approximately half the object is silver-colored metal, while the other half is covered with thick green and red corrosion products as well as some thinner gray and black corrosion. The back was crudely cleaned, leaving scratch marks in a number of areas, especially around the relief decoration. Approximately one-quarter of the reflective surface is silver-colored metal with some thin black corrosion products. The remaining surface is covered with thick green, red, and black deposits. Preserved in the corrosion products are plain-weave textile pseudomorphs, some textile fibers, hair-like pseudomorphs, and possibly some hair. Since the hair-like structure seems too coarse for human hair, it may be horsehair.

神仙山岳菱花鏡 (no. 62)
Octafoil Mirror with Riding Immortals and Rising Peaks
Tang (618–907)
Weight: 334 g
Diameter: 12.2 cm
Rim height: 0.75 cm
Overall height (rim and knob): 0.75 cm
1995.344

Manufacture: Cast in high-tin, low-lead bronze by the lost-wax method. The x-radiograph reveals one area of porosity at the edge. A chisel was used to smooth the metal in the border scallops. The back was polished, softening the relief decoration to some extent.

Condition: Much of the mirror's bright silver-gray metal is preserved. On the back are several scattered areas of raised green corrosion with some red at the edges. Approximately half of the reflective surface is covered with raised green corrosion with some light tan-colored burial dirt mixed in. Some post-burial abrasion of the reflective surface is evident.

雙鳳雙獸菱花鏡 (no. 64)
Octafoil Mirror with Paired Phoenixes and Animals
Tang (618–907)
Weight: 493 g
Diameter: 12.3 cm
Rim height: 0.8 cm
Overall height (knob): 1 cm
1995.345

Manufacture: Cast in high-tin, low-lead bronze by the lost-wax method. Extensive finishing included chiseling and polishing the flat background. Polishing of the relief decoration softened the cast details.

Condition: Much silver-colored metal is preserved, especially on the mirror back. Small areas of black and green corrosion are scattered over the back. The reflective surface shows extensive mottled black corrosion products with small areas of red and green corrosion. One of the points on the reflective side is chipped, and random post-excavation scratches are present. The black corrosion preserves plain-weave textile pseudomorphs.

雙鳳花鳥葵花鏡 (no. 67)
Lobed Mirror with Paired Phoenixes, a Nestling Bird, and a Lotus Blossom
Tang (618–907)
Weight: 942 g
Diameter: 19.1 cm
Rim height: 0.65 cm
Overall height (knob): 1.1 cm
1995.346

Manufacture: Cast in high-tin, low-lead bronze by the lost-wax method. The back is highly finished, with chisel marks around the relief decoration and at the raised border. The flat surfaces are polished, as are some in the relief decoration raised areas.

Condition: The back of the mirror is bright silver to gray-colored metal with random areas of gray-black corrosion products. Only small areas of green corrosion are observed on the knob and edges. Approximately half of the reflective surface is covered with thick green corrosion products; the remainder is the silver-gray metal. Post-burial abrasion appears on the reflective side.

禽獸纏枝菱花鏡
Hexafoil Mirror with Floral Interlaces, a Phoenix, and Running Animals
Tang (618–907)
Weight: 798 g
Diameter: 20 cm
Rim height: 0.75 cm
Overall height (knob): 1 cm
1995.347

Manufacture: Cast in high-tin, low-lead (1.5% arsenic) bronze by the lost-wax method. Extensive porosity is visible on the cast surfaces as well as in the break cross-section. The extent of porosity, corrosion, and repatination makes it impossible to determine the degree of finishing work. Fine, evenly distributed porosity can be seen in the x-radiograph.

Condition: The mirror is broken in half, with a small piece missing from the center area. Adhesive residue, perhaps an epoxy, exists on the break edges and on the back and reflective surfaces. Based on the extreme pitting and erosion of the bronze surface, the mirror appears to have been extensively corroded. Most of the corrosion has been removed—probably by chemical or electrochemical methods—and a thick black wax-like coating applied. Some areas of green corrosion are present; one such area on the back looks natural, while the remainder may have been applied.

雙鳳花枝葵花鏡 (no. 66a)
Lobed Mirror with Paired Phoenixes and Floral Displays
Tang (618–907)
Weight: 692 g
Diameter: 14.8 cm
Rim height: 0.55 cm
Overall height (knob): 1.1 cm
1995.348

Manufacture: Cast in high-tin, low-lead bronze by the lost-wax method. A chisel was used to smooth the flat metal surface around the relief decoration and next to the raised edge. The flat surfaces were polished. The cast metal details show little to no finishing work, with rough metal left in the recessed areas of the relief. Some polishing was done to the highest areas of the relief decoration.

Condition: Both sides of the mirror have a thick lustrous black to black-green surface. Several areas of raised green corrosion appear on the border of the mirror back. Much of the reflective surface is covered with light random post-burial scratches, with an area of deep post-burial scratches in the center area. There are several areas of green subsurface corrosion around the perimeter of the reflective side, along with some small corrosion pits.

雙鳳花枝葵花鏡 (no. 66b)
Lobed Mirror with Paired Phoenixes and Floral Displays
Tang (618–907)
Weight: 676 g
Diameter: 15.6 cm
Rim height: 0.6 cm
Overall height (knob, including corrosion): 0.9 cm.
1995.349

Manufacture: Cast in high-tin, low-lead bronze by the lost-wax method. Some porosity is visible on the back edge. Large pores are evident in the x-radiograph. The background on the mirror back was smoothed and polished. Chisel marks and changes in metal levels are seen around the cast decoration. The high areas of the relief decoration were also polished.

Condition: Much intact silver-gray metal is present, especially on the mirror back. Patches of red and green corrosion products are scattered on the back. The reflective surface shows patches of silver-colored metal, but much more extensive corrosion is present in the form of red and green corrosion with occasional areas of blue deposits and some wart-like corrosion pits.

雙鳳雙獸葵花鏡 (no. 65)
Lobed Mirror with Paired Phoenixes and Auspicious Animals
Tang (618–907)
Weight: 2,340 g
Diameter: 27.6 cm
Rim height: 0.8 cm
Overall height (knob): 1.2 cm
1995.350

Manufacture: Cast in high-tin, low-lead bronze by the lost-wax method. Chisel and abrasion marks on the flat areas disappear close to the relief decoration. Rough metal from casting is adjacent to the relief decoration. Polishing of the relief decoration softened the cast details.

Condition: The silver-colored metallic surface is well preserved. Approximately half of the back and 90% of the reflective surface are silver colored. The remaining back surface—mottled red and black in color—is covered with thin black and thick red, green, and blue corrosion products. Thicker green corrosion was removed with chemicals and abrasion. Tool marks appear in the corroded areas. There is also some redeposited copper, perhaps the result of chemical treatment.

雙鳳上圓下方葵花鏡 (no. 69)
Lobed Mirror with Heaven and Earth between the Paired Phoenixes
Tang (618–907)
Weight: 1,533 g
Diameter: 22.5 cm
Rim height: 0.7 cm
Overall height (knob): 1.1 cm
Inscription: 上圓下方，象于天地。中列八卦，備著陰陽。辰星鎮定，日月貞明。周流為水，以名四瀆。內置連山，以旌王岳。
1995.351

Manufacture: Cast in high-tin, low-lead bronze by the lost-wax method. The recessed areas were smoothed and polished except close to the raised decoration. The high areas of the relief decoration were also polished.

Condition: Much of the silver-colored metal is preserved. The edge of the back is mostly covered with thick red and green corrosion products, where a plain-weave textile pattern is found. One area also exhibits a felted textile pattern. The center of the back shows small scattered areas of red corrosion. The reflective surface is more heavily covered with thick red and green deposits; a thinner black corrosion layer also is present. Approximately ten chips are lost from the edge of the mirror on the reflective side; many appear to be pre-burial, as they are covered with corrosion. The post-burial chips have a much more silvery surface than the surrounding bronze.

瑞獸葡萄鏡
Animal-and-Grape Mirror
Tang (618–907)
Weight: 353 g
Diameter: 11.2 cm
Rim height: 1.05 cm
Overall height (knob): 1.3 cm
1995.352

Manufacture: Cast in high-tin, low-lead (1.5% arsenic) bronze by the lost-wax method. The decoration on the back was finished by grinding and polishing the high areas. In the less finished lower areas, some roughness of the metal is still visible.

Condition: The mirror was broken into four pieces, with one small triangular piece missing. Remains of a brown-colored adhesive (possibly shellac) are on the break edges. Much of the silver-colored bronze metal is present. Patches of red, green, and black corrosion products cover roughly half of both sides. Remains of burial dirt are present on the surface, especially the reflective side.

瑞獸葡萄鏡 (no. 55)
Animal-and-Grape Mirror
Tang (618–907)
Weight: 1,097 g
Diameter: 14.8 cm
Rim height: 1.45 cm
Overall height (rim): 1.45 cm
1995.353

Manufacture: Cast in high-tin, low-lead bronze by the lost-wax method. Very fine porosity is evenly scattered throughout the casting. Extensive finishing work was done. A marked softening of the cast features can be observed in the raised decoration.

Condition: Much uncorroded silver-colored metal remains, with scattered patches of thick red and green corrosion products on both sides. Cubic red corrosion crystals are visible on the reflective surface, and some matted textile fibers are caught in the corrosion products at the surface edge. Random light post-burial abrasion also shows on this surface.

瑞獸葡萄鏡
Animal-and-Grape Mirror
Tang (618–907)
Weight: 814 g
Diameter: 13.7 cm
Rim height: 1.4 cm
Overall height (rim): 1.4 cm
1995.354

Manufacture: Cast in high-tin, low-lead bronze by the lost-wax method. Both the relief decoration and the recessed areas on the back were abraded and polished. A marked softening of the cast features is apparent in the raised decoration.

Condition: A network of cracks runs through the metal; the larger cracks radiate from the center. Much uncorroded silver-colored metal is preserved. Thick layers of red and green corrosion products cover approximately three-quarters of the reflective surface. Wart-like corrosion pits also are present on this surface, as well as some post-burial abrasion. Approximately half of the back is covered with thick red and green deposits.

瑞獸珍禽葡萄鏡 (no. 56)
Animal-and-Grape Mirror with a Variety of Birds
Tang (618–907)
Weight: 741 g
Diameter: 13.9 cm
Rim height: 1.4 cm
Overall height (rim): 1.4 cm
1995.355

Manufacture: Cast in high-tin, low-lead (1.5 to 2% antimony) bronze by the lost-wax method. The cast metal was smoothed in both recessed and high relief areas. The high areas of relief were polished.

Condition: A thin, somewhat transparent, mottled green corrosion layer covers most of the reflective surface. Several areas of thicker light green and red-colored corrosion products also are present, along with a number of deep post-burial scratches. The mirror back remains a bright silver-gray color with several small scattered areas of thick green corrosion. Light-colored burial dirt resides in the more protected areas of the design.

瑞獸葡萄鏡 (no. 53)
Animal-and-Grape Mirror
Tang (618–907)
Weight: 215 g
Diameter: 8.2 cm
Rim height: 1 cm
Overall height (rim): 1 cm
1995.356

Manufacture: Cast in high-tin, low-lead bronze by the lost-wax method. Some finishing work is visible on the mirror back. The high areas of the decoration were smoothed and polished. The lower areas are less finished by abrasion, with a gradation to the rough pebbly surface of the cast metal in the lowest areas.

Condition: The mirror has a relatively even black patina. The glossy reflective surface contains very small (less than 1 mm in diameter) areas of wart-like corrosion pits and several deep post-burial scratches.

瑞獸天馬葡萄鏡
Mirror with Auspicious Animals, Celestial Horses, and Grapevines
Jin (1115–1234)
Weight: 2,440 g
Diameter: 25.9 cm
Rim height: 0.85 cm
Overall height (knob): 1.6 cm
1995.357

Manufacture: Cast in high-tin, low-lead bronze by the lost-wax method. Thick corrosion layers prevent determination of the extent of finishing work.

Condition: The mirror is extensively corroded, with thick layers of red and green corrosion products on both sides. Much of the corrosion was mechanically thinned from the reflective surface; a small triangular patch was left to show the original corrosion. There may be textile pseudomorphs left in the corrosion products on the reflective side, but this is difficult to determine because of the extensive cleaning. Some active chloride corrosion is present in scattered areas. The corrosion products were probably reduced with acids, as there is no evidence of mechanical tool marks.

雀繞花枝鏡 (no. 60)
Mirror with Sparrows and Floral Sprays
Tang (618–907)
Weight: 191 g
Diameter: 9.3 cm
Rim height: 0.8 cm
Overall height (knob): 0.8 cm
1995.358

Manufacture: Cast in high-tin bronze by the lost-wax method. Some finishing work in the flat background is evident. Rough cast areas are seen near the relief decoration. Raised areas of relief were polished.

Condition: Much of the mirror's surface is a metallic silver-gray color. There are random areas of thick green corrosion with smaller areas of red corrosion on the back, and larger areas of red and black corrosion on the reflective side. The reflective side was abraded to reduce the corrosion products, resulting in fine scratch marks throughout.

花卉葵花鏡 (no. 59)
Lobed Mirror with Eight Blossoms
Tang (618–907)
Weight: 1,493 g
Diameter: 24.5 cm
Rim height: 0.7 cm
Overall height (knob): 0.9 cm
1995.359

Manufacture: Cast in high-tin, low-lead (4% antimony, 3% arsenic) bronze by the lost-wax method. Extensive finishing work with a chisel is evident on the mirror back's flat background. The background was polished extensively, the high areas of the design less so. The design's recessed areas were left rough.

Condition: Most of the mirror back remains a bright silver-gray-colored metal, with several areas of thick red and green corrosion products. Except for an area where the silver-gray-colored metal remains intact, the reflective surface is covered with a thick layer of green corrosion. Magnification reveals a thick red corrosion layer under the green corrosion; some blue corrosion is also seen. A rust-colored circle, 10.5 cm in diameter, in the corrosion layer indicates that a circular or cylindrical iron object was in contact with the mirror during burial. A brown to orange-brown dirt is trapped in the corrosion.

寶相花鏡 (no. 58)
Lobed Mirror with Six Blossoms
Tang (618–907)
Weight: 961 g
Diameter: 19.1 cm
Rim height: 0.55 cm
Overall height (knob): 1.2 cm
1995.360

Manufacture: Cast in high-tin, low-lead bronze by the lost-wax method. Evidence of finishing work is visible. Chisel marks are found next to the relief decoration at the mirror's border. The background and relief decoration were polished. The low areas of the relief decoration were not finished, as some graininess of the cast metal is evident.

Condition: The mirror is well preserved with much of the silver-colored metal exposed, especially on the back. The back is covered with a layer of green and blue corrosion that is generally thin, but quite thick in some areas. The reflective surface is mostly covered with a thick layer of green, red, and black corrosion products with some blue corrosion. Much orange and brown burial mud is caked on this side of the mirror. Wart-like corrosion is present at the boundary between uncorroded metal and the heavily corroded area. Two white burial accretions appear on the edge of the reflective surface.

龜紐蓮座鏡
Mirror of Tortoise Knob and Lotus Pattern
Tang (618–907)
Weight: 373 g
Diameter: 12.5 cm
Rim height: 0.5 cm
Overall height (knob): 0.7 cm
1995.361

Manufacture: Cast in high-tin, low-lead bronze by the lost-wax method. Extensive finishing work is evident on the mirror back. Chisel marks appear at the raised border and midband and around the central relief decoration. The back was highly polished, including the high areas of the relief decoration. Little finishing work was done within the decoration's recessed areas, as the grainy metal from casting is present; this appears to be slightly reduced.

Condition: Much silver-colored metal is preserved. Areas of green, black, and red corrosion products cover approximately one-quarter of the back. About half of the reflective surface is covered with a light layer of scattered green corrosion, beneath which can be seen more extensive patches of black to gray-colored corrosion. The mirror back exhibits exceptional reflectance due to the high polish. Shallow post-burial abrasion is evident on the reflective surface's corrosion layer.

卍字鏡 (no. 75)
Quadrilobed Mirror with the 卍 Symbol
Tang (618–907)
Weight: 399 g
Dimensions: 14.1 x 14.3 cm
Rim height: 0.4 cm
Overall height (knob): 0.9 cm
1995.362

Manufacture: Cast in high-tin, low-lead bronze by the lost-wax method. Little finishing work was performed on the mirror back as raised, rough, irregular cast features are present in most areas. The raised edge was abraded smooth and polished. Little porosity is visible in the x-radiographs.

Condition: The mirror was broken into six pieces that were glued back together. Some abrasion of the metal occurred along the joins. The back is silver-colored metal with scattered areas of black to gray-black corrosion products. The reflective surface is also a silver metal color with scattered areas of black and gray corrosion products, along with patches of red, green, and blue corrosion. Burial dirt is trapped in the deposits.

三樂鏡 (no. 71)
Lobed Mirror of Three Delights
Tang (618–907)
Weight: 368 g
Diameter: 12.7 cm
Rim height: 0.55 cm
Overall height (knob): 0.9 cm
Inscription: 榮啟期問曰答孔夫子
1995.363

Manufacture: Cast in high-tin, low-lead bronze by the lost-wax method. Extensive finishing work was done to the mirror back. Abrasion is evident in the background adjacent to the raised decoration. Polishing softened the decoration.

Condition: Much of the mirror's surface is a metallic silver color. The mirror back is mostly metal colored with small scattered areas of green and black corrosion. Layers of relatively thick red and green corrosion and thin black corrosion cover approximately half of the reflective surface. A small conchoidal chip appears at the edge of this surface.

蟠龍紋鏡 (no. 73)
Mirror with Interlacing Dragons
Tang (618–907)
Weight: 621 g
Diameter: 16.9 cm
Rim height: 0.6 cm
Overall height (knob): 0.6 cm
1995.364

Manufacture: Cast in high-tin, low-lead bronze by the lost-wax method. No finishing work, not even abrasive finishing, is evident on the mirror back. Spheres and ridges of metal resulting from imperfections or cracks in the mold are scattered over the back. Chisel marks appear only on the mirror edge.

Condition: The mirror is metallic with a gray-silver appearance. A few isolated islands of green and red corrosion products are on the back. The reflective surface is mostly covered with red and green corrosion as well as burial mud. Several areas in the corrosion layers contain plain-weave textile pseudomorphs.

狩獵鏡 (no. 63)
Octafoil Mirror with Hunters and Prey
Tang (618–907)
Weight: 899 g
Diameter: 19.9 cm
Rim height: 0.75 cm
Overall height (knob): 1 cm
1995.365

Manufacture: Cast in high-tin, low-lead bronze by the lost-wax method. The back was extensively finished, with abrasion and chisel marks evident around the relief, the border, and the midband. High areas of the relief were also smoothed and polished.

Condition: Much of the silver-colored metallic surface is well preserved and visible. Both sides of the mirror contain scattered light green, black and red corrosion products. Some burial dirt is trapped in the deposits. The x-radiograph reveals a 3-cm-long blind crack extending from the base of the knob.

雙鳳花鳥葵花鏡 (no. 68)
Octafoil Mirror with Paired Phoenixes, Birds, and Flowers
Tang (618–907)
Weight: 2,740 g
Diameter: 28.8 cm
Rim height: 0.95 cm
Overall height (knob): 1.6 cm
1995.366

Manufacture: Cast in high-tin, low-lead bronze by the lost-wax method. Some porosity is evident on the mirror back and reflective surface. Abrasion marks on flat areas disappear close to the relief decoration. Rough metal from casting is adjacent to the raised sections. Circular polishing marks are visible on the reflective surface.

Condition: The silver-colored metallic surface is well preserved. A hard, dense, black corrosion product with more granular green corrosion covers approximately half of both sides. Under magnification, red corrosion can be seen in scattered areas on the surface. Abrasion marks indicate some mechanical reduction of the green corrosion from the mirror back. A large blind crack was filled and inpainted.

盤龍鏡 (no. 72)
Mirror with a Coiling Dragon
Tang (618–907)
Weight: 329 g
Diameter: 10.2 cm
Rim height: 0.8 cm
Overall height (knob): 1.1 cm
1995.367

Manufacture: Cast in high-tin, low-lead bronze by the lost-wax method. The wax appears to have been applied like cake icing to create fine detail in the relief. Possible wax drips can be observed. There is a crack in the mold across the dragon's rear proper left leg. The mirror back was carefully finished. Some ripples in the metal indicate the use of a chisel for smoothing. Subsequent polishing has virtually removed all tool marks. Little of the cast pebbled surface is visible.

Condition: A small section of the rim below the dragon's front legs was restored. Holes were drilled into the edge of the break face and a wire attached to anchor the fill. The fill is painted over with a glossy black paint. The mirror has a lustrous black patina with slight variations in the black tones. Some burial dirt and several light areas of green corrosion are found on the mirror back. Several wart-like corrosion pits appear in the reflective surface, along with some relatively shallow post-burial abrasion.

卍字鏡 (no. 76)
Mirror with the 卍 Symbol
Liao (916–1125), the Taiping reign (1020–30)
Weight: 318 g
Dimensions: 12.5 x 12.5 cm
Rim height: 0.45 cm
Overall height (knob): 0.9 cm
Inscription: 太平萬歲
1995.368

Manufacture: Cast in high-tin, low-lead bronze by the lost-wax method. Finishing work is minimal. A chisel may have been used to smooth the top of some of the swastika lines, as several areas of chattering marks are visible. The border appears to be abraded smooth. The recessed area is rough and irregular from the casting process.

Condition: Several areas of silver-colored metal are exposed. Most of the surface is covered with black, red, and green corrosion products that tend to be relatively thick, especially on the reflective side. These corrosion layers preserve plain-weave textile pseudomorphs and fibers. Hair-like pseudomorphs cover much of the reflective surface, which also is scattered with preserved textile fibers. In one corner of the same surface, a deep scratch is visible in the corrosion layers.

真子飛霜鏡 (no. 70)
Lobed Mirror with a Tortoise Knob, a Musician, and a Phoenix
Tang (618–907)
Weight: 1,394 g
Diameter: 21.6 cm
Rim height: 0.6 cm
Overall height (knob): 1.2 cm
Inscription: 鳳凰雙鏡南金裝。陰陽各為配，日月恒相會。白玉芙蓉匣，翠羽瓊瑤 帶。同心人，心相親，照心照膽保千春。
1995.369

Manufacture: Cast in high-tin, low-lead bronze by the lost-wax method. Chiseling and grinding marks from finishing work can be seen on the mirror back, especially around the raised decoration. The background area was polished.

Condition: Much of the silver-colored metal is preserved. Both sides have scattered small areas of red and green corrosion. Wart-like corrosion pits are on the reflective side.

湖州素鏡 (no. 90)
Huzhou Mirror
Ming (1368–1644)
Weight: 708 g
Diameter: 15.8 cm
Rim height: 0.9 cm
Overall height (knob): 1.1 cm
Inscription: 湖州近河造
1995.370

Manufacture: Cast in high-tin, low-lead (3% antimony, 2% arsenic) bronze by the lost-wax method. An iron chaplet is on the reflective side, and porosity is evident on the mirror back and in the x-radiographs. The back was extensively finished and polished. Abrasion marks exist around the center knob and near the raised border.

Condition: The mirror is almost entirely metallic with a thick oxide layer that varies in color from light gray to dark gray. Magnification shows some green corrosion. Random scratches are seen throughout, especially on the reflective surface. The mirror edge contains a small dent.

花瓣連錢紋方鏡
Square Mirror with Floral and Coin Motifs
Song (960–1279) or Liao (907–1125)
Weight: 100.5 g
Dimensions: 10.5 x 10.5 cm
Rim height: 0.25 cm
Overall height (knob): 0.5 cm
1995.371

Manufacture: Cast in high-tin, low-lead bronze by the lost-wax method. An abraded edge shows a yellow interior and a tin-colored surface. Several areas of the design elements did not cast properly and small sections of lines are missing. The x-radiograph reveals some porosity along one edge. Little finishing work is evident on the mirror back.

Condition: Much of the silver-colored metal is preserved on the back, with some scattered green, red, and black corrosion products. Most of the reflective surface is covered with black, red, and green corrosion to which burial dirt adheres; the small amount of uncorroded surface is a shiny metallic color.

吉祥銘文方鏡

Square Mirror with Felicitous Message
Song (960–1279)
Weight: 704 g
Dimensions: 16.3 x 16 cm
Rim height: 0.5 cm
Overall height (knob): 0.7 cm
Inscription: 長命富貴 ([May you enjoy] a long life, health, and fortune.)
1995.372

Manufacture: Cast in high-tin, low-lead bronze by the lost-wax method. Several large areas of porosity are scattered in the casting. Corrosion makes it difficult to determine the extent of finishing work on the mirror back. Some rough abrasion appears on the edges.

Condition: The mirror is extensively corroded. Covered with thick layers of green, red, and black corrosion products, the mirror back shows a scattering of burial mud and a possible rust stain near the knob. On the reflective surface, some silver-colored metal is visible through the scattered green, blue, black, and red corrosion products.

雙獅亞形鏡 (no. 74)

Cruciform Mirror with Two Lions
Tang (618–907)
Weight: 214.5 g
Dimensions: 10.4 x 10.5 cm
Rim height: 0.35 cm
Overall height (knob): 0.7 cm
1995.373

Manufacture: Cast in high-tin, low-lead bronze by the lost-wax method. The wax model apparently was created in three sections: the center decoration, the knob, and the border. The center decoration was cast in a mold, then joined to the border. The knob appears to be attached to the decoration. The circular background elements seem to have been created in the mold, perhaps with a tube. Some finishing work is evident, as the cast details were softened by abrasion and polishing. The border was polished.

Condition: Much silver-colored metal is preserved. Black, green, and some red corrosion products cover approximately half of the back. Green, black, and red corrosion covers about one-third of the reflective surface. A number of deep post-burial scratches on the reflective side abraded some of the corrosion products. One corner is distorted due to subsurface corrosion. There is a 2.2-cm-long crack in another corner.

雙鳳花枝方鏡 (no. 77)

Square Mirror with Two Phoenixes and Floral Sprays
Northern Song (960–1127)
Weight: 723 g
Dimensions: 16.2 x 16.2 cm
Rim height: 0.5 cm
Overall height (knob): 0.8 cm
1995.374

Manufacture: Cast in high-tin, low-lead bronze by the lost-wax method. A flat area on the knob suggests that a sprue may have been cut from it. An iron chaplet is visible on the reflective surface in one corner of the mirror. Moderate abrasive finishing work is evident on the flat border, and the step to the raised border shows rough-cast metal. Little to no finishing work was done in the relief decoration.

Condition: The mirror back has a flat black patina. Recessed areas around the relief decoration trap light cream-colored burial dirt. Magnification reveals some thin light green and blue corrosion on the back. The reflective surface—a glossy mottled black, gray, and green color—shows light post-burial abrasion.

月宮鏡 (no. 83)

Octafoil Mirror with Lunar Palace
Jin (1115–1234)
Weight: 954 g
Diameter: 21.3 cm
Rim height: 0.75 cm
Overall height (knob): 0.9 cm
1995.375

Manufacture: Cast in high-tin, low-lead bronze by the lost-wax method. Porosity is visible on the surface and in the x-radiograph. A fair amount of finishing work was done to the mirror back. Chisel and scrape marks exist in the background around the relief decoration. The smooth quality of the patina suggests that the back was finished using abrasion and perhaps polishing.

Condition: The mirror back is covered with an even olive green to black patina, while the reflective surface is primarily a glossy black with scattered corrosion pits. Virtually no green corrosion is present in the pits, and possibly was removed. A large irregular patch of red corrosion in the center of the reflective surface was reduced by abrasion. The mirror edge contains one area of uncorroded metal, along with a gauze textile pseudomorph in the nearby black corrosion layer.

佽非斬蛟龍葵花鏡 (no. 80)

Octafoil Mirror Featuring Ci Fei, the Dragon Slayer
Song (960–1279) or Jin (1115–1234)
Weight: 380.5 g
Diameter: 17.2 cm
Rim height: 0.45 cm
Overall height (knob): 1.1 cm
Inscription: 煌丕昌天
1995.376

Manufacture: Cast in high-tin, low-lead bronze by the lost-wax method. Extensive large voids due to porosity are visible in the x-radiograph. The degree of finishing work is difficult to determine because of corrosion. Based on the smooth patina, it is likely that abrasion and polishing were done.

Condition: The mirror surface is completely mineralized in a compact, slightly mottled olive-colored patina typical of high-tin bronzes, with several scattered small areas of green corrosion. The brittle nature of the patina caused the edge of the mirror to chip. The reflective surface contains several post-burial scratches, one of which is long and deep.

單龍鏡

Dragon Mirror
Ming (1368–1644)
Weight: 739 g
Diameter: 14.1 cm
Rim height: 0.95 cm
Overall height (knob): 1.3 cm
1995.377

Manufacture: Cast in high-tin, low-lead bronze by the lost-wax method. Cleaning of the mirror, which was once heavily covered with corrosion products, makes it difficult to determine the extent of finishing work. Even so, the back appears to have been carefully finished, and the border polished.

Condition: The mirror surface is essentially metallic with a dark gray to black color. Patches of red and green corrosion are scattered on the back. Magnification reveals scratches and abrasion marks in the relief decoration, indicating that many of the corrosion products were mechanically removed. On the reflective surface there is much green and red corrosion, with small areas of blue corrosion. Textile pseudomorphs and some textile fibers are preserved in the corrosion products. Below the dragon is a blind crack, visible only in the x-radiograph.

渡海二仙菱花鏡 (no. 82)

Octafoil Mirror with Two Immortals Crossing the Ocean
Song (960–1279)
Weight: 606 g
Diameter: 17.6 cm
Rim height: 0.6 cm
Overall height (knob): 1 cm
Inscription: 馳名周原造
1995.378

Manufacture: Cast in low-tin, high-lead bronze by the lost-wax method. A fair amount of porosity is visible in the x-radiograph. The degree of finishing work cannot be determined because of extensive corrosion.

Condition: The mirror surface is covered with corrosion products, primarily green, red, and black. On the reflective side, a number of deep scratches are covered with corrosion and appear to have occurred before burial. Many more post-burial abrasion marks apparently resulted from surface corrosion cleaning. The mirror back contains burial dirt. One large chip and several smaller chips are seen in the corrosion layer around the mirror edge.

雙魚鏡 (no. 86)

Mirror with Twin Carp
Jin (1115–1234)
Weight: 813 g
Diameter: 17.9 cm
Rim height: 0.7 cm
Overall height (knob): 0.9 cm
1995.379

Manufacture: Cast in moderate-tin, high-lead bronze by the lost-wax method. Extensive porosity is visible in the x-radiograph. Corrosion and abrasion make it difficult to determine the extent of finishing work.

Condition: A relatively thick mixture of red, green, and black corrosion covers the mirror. On the reflective surface the green corrosion is often spherical or in the form of broken spheres. The corrosion products were partially removed from the relief decoration by abrasion, leaving scratches in the metal. Two areas of deep post-burial scratches are present on the reflective side. Around the back rim are a number of pits of active corrosion.

沙羅樹鏡

Lobed Mirror with *Shaluo* Tree
Northern Song (960–1127)
Weight: 265 g
Diameter: 14.7 cm
Rim height: 0.4 cm
Overall height (knob): 0.8 cm
1995.380

Manufacture: Cast in low-tin, high-lead bronze by the lost-wax method. Some surface porosity is visible, and the x-radiographs reveal scattered medium-size porosity. A network of recessed lines in the mirror back may have been caused by cracks in the wax model. Extensive abrasive finishing work was performed. The background is smooth and free of casting artifacts, and the relief decoration is softened.

Condition: A gray-colored corrosion covers much of the surface, with occasional patches of metal. Small areas of red, green, and blue corrosion are found on the reflective surface. The back has extensive thin blue corrosion products and small islands of blue and green glassy spheres of corrosion—much of which may be chemically applied. The back mirror edge contains a large rust stain.

大定通寶花紋鏡 (no. 84)
Mirror Featuring *Dading* Coins
Jin (1115–1234)
Weight: 139 g
Diameter: 11.9 cm
Rim height: 0.4 cm
Overall height (knob): 0.8 cm
Inscription: 大定通寶
1995.381

Manufacture: Cast in low-tin, high-lead bronze by the lost-wax method. A metal ridge and a clump of metal near the edge formed due to a crack and damage in the mold. The edge shows some porosity. There is little evidence of abrasive finishing work.

Condition: The mirror is extensively corroded, with both sides covered with thick layers of black, red, and green products. Several dents appear in the rim.

童子騎馬鏡 (no. 85)
Mirror with Wheel Spokes and Riding Boys
Jin (1115–1234)
Weight: 311 g
Diameter: 13.3 cm
Rim height: 0.7 cm
Overall height (knob): 0.8 cm
Inscription: 官立鏡子局
1995.382

Manufacture: Cast in moderate-tin, high-lead bronze by the lost-wax method. Some finishing work is evident. The background is smoother than the areas immediately adjacent to the relief decoration. The raised decoration is very worn, possibly from polishing and/or subsequent wear.

Condition: A black to brown patina covers the mirror. Above the black layer lie patches of thin light green corrosion products, interspersed with light-colored burial mud. Deep ragged scratches appear in the reflective surface, revealing a copper-colored metal. The relief decoration on the mirror back, the fine details of which have been worn away, contains some abrasion and scratches.

牛郎織女菱花鏡 (no. 81)
Octafoil Mirror with Legend of Herdboy and Weaver Maid
Song (960–1279)
Weight: 638 g
Diameter: 17.8 cm
Rim height: 0.7 cm
Overall height (knob): 1.1 cm
1995.383

Manufacture: Cast in moderate-tin, high-lead bronze by the lost-wax method. Much porosity is visible in the x-radiograph along with some surface porosity, especially in the rim at the bottom. X-radiographs also reveal a round repair in the robes of the main figure on the mirror back's left side. The extent of finishing work is difficult to determine due to extensive corrosion.

Condition: Extensive surface corrosion includes black, green, and red deposits on both sides. The corrosion products give the mirror a mottled appearance with extensive erosion and surface loss. Some burial mud is trapped in the mirror back's corrosion layers.

畫像帶柄鏡
Narrative Mirror with Handle
Jin (1115–1234) or Yuan (1279–1368)
Weight: 423 g
Dimensions: L. 18.2 cm, diam. 9.7 cm
Rim height: 0.5 cm
Overall height (rim): 0.5 cm
1995.384

Manufacture: Cast in moderate-tin, high-lead bronze by the lost-wax method. At the back center of the handle is an iron chaplet; the x-radiograph reveals a second chaplet higher up on the handle. The mirror is very worn, making it difficult to assess accurately the degree of finishing. The high areas of the design appear to have been polished, with little finishing work on the background, where the metal is still rough.

Condition: The mirror surface is now a black color covered with green and red corrosion products—the result of a process that has caused some pitting and metal erosion. Corrosion products on the reflective surface have been removed by abrasion, leaving deep parallel scratches in the metal.

Mirror Stand in the Shape of a Djeiran
Qing (1644–1911)
Weight: 1,162 g
Dimensions: 21.2 x 11.1 x 10.6 cm
1995.385

Manufacture: Cast in low-tin, low-lead bronze by the lost-wax method, in two sections: the animal and the mirror bracket. X-radiographs reveal a number of chaplets, several nubs of which are inside the animal. Porosity is present near the animal's bottom edge and in three areas of the bracket. The bracket is held in place by a rivet and the remains of what appears to be a lead-tin solder around the join. The bronze surface was highly finished by grinding and polishing. Chasing tools were used to sharpen the cast details.

Condition: The bracket moves on the animal's back but is secure. The bronze surface is covered with black and green corrosion products with areas of red corrosion, along with a black resinous material mixed with dirt.

Mirror Stand in the Shape of an Ox
Qing (1644–1911)
Weight: 817 g
Dimensions: 14.2 x 7.8 x 9.6 cm
1995.386

Manufacture: Cast in low-tin, low-lead bronze (containing 4% zinc) by the lost-wax method, in one piece. The casting is riddled with porosity, visible in the x-radiograph. The surface was finished by grinding and polishing, not as finely done near the areas of relief. The details were chased.

Condition: The stand has an even red to black-colored patina, with several small areas of green corrosion. Dust and dirt cling to the recessed areas. A white accretion line is visible inside the animal.

三龍鏡
Mirror with Three Dragons
Jin (1115–1234)
Weight: 329 g
Diameter: 10.2
Rim height: 0.8 cm
Overall height (knob): 1 cm
Inscription: 官 (Official), plus an unidentified cipher
1995.387

Manufacture: Cast in high-tin, moderate-lead bronze by the lost-wax method. Some surface porosity can be seen. Some finishing work is on the mirror back, with abrasive marks on the stepped surfaces and adjacent areas. Softening of the relief decoration is evident. The top of the knob is flat.

Condition: Black, green, and red corrosion products cover the mirror. On the reflective surface a small area of a plain-weave textile pseudomorph is preserved. A 5.5-cm-long crack extends from the middle to the edge, with small chip losses visible along its length.

雙龍荷花鏡 (no. 87)
Mirror with Twin Dragons and Lotus Blossoms
Yuan (1279–1368), dated 1338
Weight: 1,676 g
Diameter: 23.4 cm
Rim height: 0.8 cm
Overall height (knob): 1.1 cm
Inscription: 至元四年，河中府馬溝玉家自家門人監照白銅鏡。
1995.388

Manufacture: Cast in moderate-tin, moderate-lead bronze by the lost-wax method. Rough islands of metal and no softening of cast detail indicate that little finishing work was done to the mirror back. The top of the center knob is flat. The only large areas of porosity are at one edge.

Condition: Extensive black corrosion products, interspersed with areas of green corrosion, cover both sides of the mirror. The green corrosion is thicker on the mirror back, which also contains some burial mud. The reflective side shows many random deep scratches, most of them patinated.

雙鳳鏡 (no. 79)
Mirror with a Pair of Phoenixes
Song (960–1279)
Weight: 3,760 g
Diameter: 35.1 cm
Rim height: 0.75 cm
Overall height (knob): 1.3 cm
Inscription: 金陵張子皋
1995.389

Manufacture: Cast in high-tin, high-lead bronze by the lost-wax method. The degree of finishing work is difficult to determine due to extensive corrosion. The background areas were probably abraded smooth and polished.

Condition: An assortment of green, red, black, and blue corrosion products cover the mirror. Some of the blue and green corrosion appears to have been chemically induced. The reflective surface was abraded to remove corrosion products, leaving deep scratches in the metal and corrosion products. Remains of a plain-weave textile pseudomorph are preserved in some of the thicker corrosion patches on the reflective surface.

雙龍鏡
Mirror with Two Dragons
Date unknown
Weight: 407 g
Diameter: 9.9 cm
Rim height: 1.1 cm
Overall height (knob): 1.8 cm
1995.390

Manufacture: Cast in brass by the lost-wax method, in two sections. The central knob on the back was cast separately and set into a hole, around which metal was then tooled to hold the knob in place. A decorative set of punch marks surrounds the perimeter of the hole. Much of the relief decoration was finished with tooling or chasing. The linear patterns were chased.

Condition: Thin layers of black and green corrosion cover the mirror's surface. Areas of red corrosion products and possible copper chloride corrosion also are visible. Part of the mirror back contains rust stains and small particles of iron. The knob is not locked in place by corrosion and moves freely. A blind crack near the edge is visible only in the x-radiographs.

春宮鏡
Mirror with Erotic Scenes
Yuan (1279–1368)
Weight: 84.5 g
Diameter: 7.2 cm
Rim height: 1 cm
Overall height (rim): 1 cm
1995.391

Manufacture: Cast in high-tin, moderate-lead bronze by the lost-wax method. Cast features are worn, apparently the result of abrasive finishing.

Condition: Thick black and green corrosion products cover the mirror back, with many low areas trapping orange-colored burial mud and a white-colored accretion. Mottled green, black, and gray-colored corrosion products cover the reflective surface, which also shows some white burial accretions and orange burial dirt.

瑞獸葡萄鏡
Animal-and-Grape Mirror
Jin (1115–1234)
Weight: 562 g
Diameter: 12.1 cm
Rim height: 1.15 cm
Overall height (knob): 1.2 cm
1995.392

Manufacture: Cast in moderate-tin, low-lead bronze by the lost-wax method. Relief decoration was finished by moderate abrasion with perhaps some chased details. Little finishing work is evident in the background areas.

Condition: The mirror is metallic with a dark brown to black patina. On the back are dust, dirt, and a light green corrosion. The patina on the reflective side is two-toned: a smooth glossy black color in the center area, and a smooth glossy red-brown color on the perimeter. Random post-burial scratches and dents cover the reflective surface, and the back edge contains two metal chip losses.

四乳四虺鳥紋鏡 (no. 91)
Mirror with Four Nipples, Four Quasi-Dragons, and Birds
Ming (1368–1644)
Weight: 355 g
Diameter: 10.5 cm
Rim height: 0.75 cm
Overall height (knob): 1.1 cm
Inscription:宮
1995.393

Manufacture: Cast in high-tin, moderate-lead bronze by the lost-wax method. The x-radiograph reveals much porosity. The linear decoration lacks sharpness and crispness, and little finishing work is evident in the raised decoration. The raised inscription appears to be finished with tooling.

Condition: A thin black to brown patina covers the mirror. The back contains an area of green corrosion with three white spots of deposited material and associated light green corrosion. The reflective surface is more silvery in color with isolated black spots.

花鳥紋菱花鏡
Octafoil Mirror with Avian and Floral Motifs
Five Dynasties (907–960)
Weight: 155 g
Diameter: 13.8 cm
Rim height: 0.3 cm
Overall height (knob): 0.7 cm
1995.394

Manufacture: Cast in high-tin, low-lead bronze by the lost-wax method. Corrosion makes it difficult to determine the extent of finishing work.

Condition: Mottled olive green, black, and light green corrosion products cover the mirror, some areas of which are almost completely mineralized and brittle. Three holes have opened where the metal is thin. At the edge close to the smallest hole is an area with several blind cracks. Also near the edge are small chips in the brittle corrosion. Some burial dirt clings to the surface.

為善最樂鏡 (no. 89)
Mirror with Confucian Maxim
Ming (1368–1644)
Weight: 71.5 g
Diameter: 8.3 cm
Rim height: 0.45 cm
Overall height (knob): 0.6 cm
Inscription: 為善最樂
1995.395

Manufacture: Cast in moderate-tin, low-lead bronze by the lost-wax method. Some surface porosity is present. Extensive corrosion makes it impossible to determine the degree of finishing. It appears that the back was polished.

Condition: Green, black, and blue corrosion products cover the mirror back. A diagonal scratch runs between the two sets of characters. The reflective surface was abraded, revealing a bronze-colored metal. Green and black corrosion products are thickest near the edge, where the reflective surface is lowest. The rim is slightly bent and distorted.

雙喜五福帶柄鏡 (no. 92)
Mirror with Handle, Decorated with "Double Happiness" and Five Bats
Qing (1644–1911)
Weight: 310.5 g
Dimensions: L. 26.7 cm, diam. 16.5 cm
Rim height: 0.6 cm
Overall height (rim): 0.6 cm
Inscription: 囍。湯萬亨。
1995.401

Manufacture: Cast in brass by the lost-wax method. The design's high areas are polished and the background was left rough. Remains of a black resinous material are on the back.

Condition: The mirror contains oxides and dirt as well as scattered white accretions on the back. One spot of green corrosion is present on the rim at the 7 o'clock position. The reflective surface and the handle on that side were brightly polished and are now slightly oxidized; the surface shows random scratches.

金銀平脫花鳥鏡 (no. 61)
Mirror with Flowers and Birds in Gold and Silver Cutouts over a Lacquer Base
Tang (618–907)
Weight: 377 g
Diameter: 16.8 cm
Rim height: 0.35 cm
Overall height (knob): 0.9 cm
1999.217

Manufacture: Cast in high-tin, low-lead bronze by the lost-wax method. The center knob and edge band appear to have been finished, but the extent of this work is difficult to determine because of thick corrosion. Foliage, birds, and ribbons were inlaid in gold and silver on the back, with the details chased and punched. Lacquer probably once surrounded the inlays, but no evidence of it remains.

Condition: Extensive thick green, red, and black corrosion products cover the mirror, and a white burial accretion is present on the reflective surface. This accretion and the other corrosion products contain plain-weave textile pseudomorphs. The silver on the back is tarnished to a dark gray to black color. There is also some tarnish on the gold.

葡萄紋掌中鏡 (no. 54)
Miniature Mirror with Grape Decoration
Tang (618–907)
Weight: 61 g
Diameter: 4.9 cm
Overall height (rim): 0.7 cm
1999.218

Manufacture: Cast in high-tin, low-lead bronze by the lost-wax method. Except for some abrasion on the top of the center knob, there is little evidence of finishing work. Otherwise, rough metal is left from the casting process. An air bubble on the rim of the back was abraded, creating a recessed area.

Condition: The mirror, which has a gray metallic appearance, is corrosion free except for several small spots of green corrosion on the back and several areas of black and green corrosion on the reflective surface. Several wart-like corrosion pits are forming on the edge of the reflective surface. Much random post-burial abrasion is visible on the reflective side.

方形掌中鏡 (no. 48)
Miniature Square Mirror
Sui (581–618)
Weight: 42 g
Dimensions: 3.8 x 3.8 cm
Rim height: 0.5 cm
Overall height (knob): 0.6 cm
1999.219

Manufacture: Cast in high-tin, low-lead bronze by the lost-wax method. Little finishing work is evident on the back. The edge shows some rough abrasion marks. Otherwise, the rough metal from the casting process remains.

Condition: Much preserved silver to silver-gray metal can be seen on both sides of the mirror, along with green, red, black, and blue corrosion products. There is some wart-like corrosion in one corner of the reflective surface and some erosion and loss of metal on the back edge. A parallel linear pattern appears in the black corrosion on the reflective surface.

透雕紋鏡 (no. 4)
Double-tier Mirror
Warring States (475–221 BC)
Weight: 110 g
Diameter: 7.6 cm
Rim height: 0.5 cm
Overall height (knob): 0.8 cm
1999.220

Manufacture: The mirror was cast in two sections in two ceramic molds. The reflective surface is cast in high-tin, low-lead bronze; the back is cast in a low-tin, high-lead alloy. The method used to attach the two sections together is not known.

Condition: The reflective piece appears to have been removed to clean corrosion from the pierced areas of the mirror back. The back of the reflective piece is severely abraded. Thick green and red corrosion products cover the reflective surface; some silver metal is visible. The mirror back is completely covered with red and green corrosion layers, which were thinned by mechanical cleaning.

蓮花鏡 (no. 88)
Mirror with Lotus and Mantra
Yuan (1279–1368)
Weight: 787 g
Diameter: 22.3 cm
Rim height: 0.7 cm
Overall height (rim and convex reflective surface): 1.5 cm
1999.221

Manufacture: Cast in low-tin, high-lead bronze by the lost-wax method. X-radiographs reveal some scattered porosity. Little to no finishing work is evident on the back.

Condition: Green, red, and black corrosion products cover the reflective surface, with some light-colored burial dirt present near the edges. Some of the green corrosion is abraded to reveal the black corrosion layer beneath. The mirror back, covered with green and black corrosion products, has much light-colored burial mud trapped in the design. The knob is damaged, with half of the metal at the top missing.

亞字形纏枝六花鏡 (no. 78)
Cruciform Mirror with Six Blossoms
Northern Song (960–1127)
Weight: 408 g
Dimensions: 16.5 x 16.5 cm
Rim height: 0.35 cm
Overall height (knob): 0.7 cm
1999.222

Manufacture: Cast in high-tin, low-lead bronze by the lost-wax method. The design around the knob did not cast properly. X-radiographs reveal five round circles of less dense material that are plug repairs made to correct surface defects. Two are filled in with a yellow-colored metal, two others are seen as holes. Finishing work is evident on the back. Rough rotary marks show on the center knob. A chisel was used to smooth the background between the raised edge and the circle containing the floral decoration. The raised border was polished. Little work was done in the relief area, as there are several ridges of metal from cracks in the mold.

Condition: The reflective surface is covered with black corrosion and mottled green and red corrosion products; much light-colored burial dirt also is present. The back has an even black corrosion layer with patches of green corrosion on the raised edge. Plain-weave textile pseudomorphs are visible in the green corrosion layer.

Mirrors from Japan

八乳博局式鏡
TLV Mirror with Eight Nipples
4th–5th century
Weight: 216 g
Diameter: 12.3 cm
Rim height: 0.35 cm
Overall height (knob): 0.8 cm
1995.293

Manufacture: Cast in high-tin, high-lead bronze in a clay mold (?). Chill flow lines are visible on the mirror back. The quality of casting and finishing is not typical of Eastern Han mirrors. Much surface porosity can be seen. There is no evidence of chisel work in the T, L, and V channels, nor is there any rotary abrasion of the nipples and center knob.

Martha Carter bought this mirror from a dealer in Japanese art and suspects that it may be a Japanese recast. X-ray fluorescence analysis was carried out and the data analyzed in the same manner as Masaaki Sawada's research on Chinese and Japanese mirrors. Sawada showed that mirrors made in China and Japan could be separated by plotting the Sn-Kβ/Cu-Kβ peak counts against the Pb-Lα/Cu-Kβ peak counts. Mirrors having a high ratio of tin to lead and a slope of 1.1 or greater were Chinese in origin. Mirrors having a high ratio of lead to tin and a slope of 0.7 or less were Japanese in origin. The origin of mirrors whose slopes were between 1.1 and 0.7 could not be determined. The x-ray fluorescence data were taken from 15 locations and the Sn-Kβ/Cu-Kβ peak counts plotted against the Pb-Lα/Cu-Kβ peak counts. The line having the best fit with the data had a 0.3 slope, which is well below the 0.7 threshold for Japanese mirrors determined by Sawada. In general, the reflective surface contained a higher proportion of tin.

Condition: Thick green, red, and black corrosion products cover both sides of the mirror. On the back, the corrosion process has softened some of the cast details. Some loss of material also has occurred on the edge. Specular light shows a faint textile impression in the corrosion layers on the mirror's reflective side.

松竹雙鶴鏡
Mirror with Pine, Bamboo, and Cranes
Edo (1615–1868)
Weight: 856 g
Diameter: 11.8 cm
Rim height: 1.8 cm
Overall height (rim): 1.8 cm
1995.396

Manufacture: Cast in high-tin, low-lead bronze by the lost-wax method. The mirror back shows some surface porosity, and its rim was abraded and polished. Other areas received a minimal amount of finishing work.

Condition: Except for a few scattered areas of green corrosion on the back, the mirror is entirely metallic. A number of small holes in the silver-colored layer reveal the copper-colored alloy beneath. The reflective surface has a number of random scratches.

松鶴鏡
Mirror with Pine and Cranes
Edo (1615–1868)
Weight: 291 g
Diameter: 10.5 cm
Rim height: 1.03 cm
Overall height (rim): 1.03 cm
Signed: 天下一
1995.397

Manufacture: Cast in low-tin, low-lead bronze by the lost-wax method. The x-radiograph reveals some porosity. Some finishing work was performed on high areas of the relief decoration, but there is little evidence of other finishing.

Condition: The mirror is entirely metallic. The back, which is dusty and dirty, has a dark brown to black patina. The reflective surface has a warm oxidized copper-like patina, along with a number of random dents, scratches, and abrasion marks.

龜紐雙鵲鏡
Mirror with a Tortoise Knob and Magpies
Edo (1615–1868)
Weight: 432 g
Diameter: 11.5 cm
Rim height: 1.2 cm
Overall height (rim): 1.2 cm
1995.398

Manufacture: Cast in copper by the lost-wax process. X-radiographs reveal scattered areas of relatively large air bubbles. The mirror back contains the remains of a resinous black material.

Condition: Entirely metallic, the mirror has a dark black to brown patina. The center of the reflective side contains numerous dents. Adjacent to the dents is an area of fresh deep scratches. Some dirt and dust are on the mirror back, along with a few scattered patches of green corrosion.

松竹雙鶴柄鏡
Mirror with Handle, Decorated with Pine Tree and Crane
Edo (1615–1868)
Weight: 352 g
Dimensions: L. 25.6 cm, diam. 16.1 cm
Rim height: 0.4 cm
Overall height (rim): 0.4 cm
Signed: 滕原光正作
1995.399

Manufacture: Cast in copper with antimony (8%) and arsenic (2%) by the lost-wax method. The reflective surface has a layer of mercury. A black resinous material on the mirror back makes it difficult to determine the degree of finishing work. High areas of the relief were smoothed. Also, some chasing work probably was done to sharpen the cast decoration.

Condition: The mirror is entirely metallic with only brown and black oxides on the surface. Its handle shows an accumulation of grime mixed with a little green corrosion.

蝴蝶牡丹柄鏡

Mirror with Handle, Decorated with a Butterfly and Blooming Peony
Edo (1615–1868)
Weight: 568 g
Dimensions: L. 24.2 cm, diam. 14.5 cm
Rim height: 0.5 cm
Overall height (rim): 0.5 cm
Inscription: 天下一橋本正辛
1995.400

Manufacture: Cast in brass with tin (5.5%) and lead (4%) by the lost-wax method. The x-radiograph reveals scattered porosity. High areas were smoothed and polished; there is little evidence of finishing in the design's low areas. The mirror back may have the remains of a black lacquer or resinous coating.

Condition: The metal is well preserved, with only some surface oxidation, dirt, and grime. Several small areas of green corrosion appear on the mirror back, and the reflective surface has a number of small dents in its center. A small crack on the reflective side has formed between the mirror and handle. A small piece of the rim has broken off near the top of the mirror.

Mirrors from the Near East

Mirror with Addorsed Sphinxes
Eastern Anatolia or Northern Mesopotamia?
13th century?
Weight: 170 g
Diameter: 10.5 cm
Rim height: 0.4 cm
Overall height (knob): 0.9 cm
1995.402

Manufacture: Cast in brass by the lost-wax technique. Porosity is visible on the surface. The mirror back shows extensive finishing work. Details were chased and the background was worked with a chasing tool. Dots on the wings and circles on the tails were punched. The reflective surface is uneven—the result of the casting process—and may not have been smoothed and polished. One spot may be a sprue; other areas appear unfinished.

Condition: Black and red corrosion products cover the mirror, with deep post-burial abrasion found throughout. Some abrasion on the back exposes the yellow-colored metal. With magnification, a small crack can be seen at the edge.

Mirror with Four Leaves
Eastern Anatolia or Northern Mesopotamia?
13th century?
Weight: 69 g
Diameter: 7.7 cm
Rim height: 0.3 cm
Overall height (knob): 0.9 cm
1995.403

Manufacture: Cast in high-tin, low-lead bronze by the lost-wax method. The decoration has a soft appearance as if wax had been dripped across the surface. The hole in the center knob appears to have been formed by piercing a soft material. The back was roughly scraped between the raised decorations.

Condition: The mirror's black patina is compact and smooth. Small scattered pits of green corrosion appear on the reflective surface. Mud with some green corrosion is scattered on the back.

Bibliography

Anji 1996
Anji County Museum. "Excavation of the Western Han Tombs on Shangmashan Hill, Anji County, Zhejiang." *Kaogu* 考古 7 (1996), 46–59.

Bai 1999
Bai Yunxiang. "A Study of Bronze Mirrors with the Inscription 'Jian Ri Zhi Guang Tian Xia Da Ming' and Grass-leaf Designs and Their Molds in the Western Han Period." *Kaogu* 4 (1999), 65–78.

Ba'youwen/Chengshun 1980
Ba'youwen and Chengshun. "A Hoard of Silver of the Liao Dynasty Discovered at Balin Youqi in Ju Ud Meng, in Nei Monggol." *Wenwu* 文物 5 (1980), 45–51.

Beijing 1963
Tang yong tai gong zhu mu bi hua 唐永泰公主墓壁畫. Beijing, 1963.

Bian 1977
Bian Xiaoxuan 卞孝萱. "Handicrafts in Yangzhou during the Tang Dynasty and Excavated Relics 唐代揚州手工業與出土文物." *Wenwu* 9 (1977), 31–37.

Bulling/Drew 1971–72
A. Gutkind Bulling and Isabella Drew. "The Dating of Chinese Bronze Mirrors." *Archives of the Chinese Art Society of America* XXV (1971–72), 36–57.

Cahill 1986
Suzanne Cahill. "The Word Made Bronze: Inscriptions on Medieval Chinese Bronze Mirrors." *Archives of Asian Art* XXXIX (1986), 62–70.

Cammann 1948
Schuyler Cammann. "The TLV-pattern on Cosmic Mirror of the Han Dynasty." *Journal of the American Oriental Society* 68 (October–December 1948), 159–68.

Chase 1983
W. T. Chase. "Bronze Casting in China: A Short Technical History." *The Great Bronze Age of China: A Symposium.* Edited by George Kuwayamei. Seattle and London, 1983.

Chase/Franklin 1979
W. T. Chase and Ursula Martius Franklin. "Early Chinese Black Mirrors and Pattern-Etched Weapons." *Ars Orientalis* XI (1979), 215–58.

Chen B 1982
Chen Benming 陳本明. "Yunnan zhao tong ci ni ba chu tu liang mian tong jing 雲南昭通茨泥巴出土兩面銅鏡." *Kaogu* 3 (1982), 332.

Chen GH 1987
Chen Gaohua 陳高華. *Sui Tang hua jia shi liao* 隋唐畫家史料. Beijing, 1987.

Chen P 1987
Chen Peifen 陳佩芬. *Shanghai bo wu guan cang qing tong jing* 上海博物館藏青銅鏡. Shanghai, 1987.

Cheng TK 1948
Cheng Te-kun. "Three Dated Chinese Mirrors in the West China Union University Museum." *Oriental Art* I (2) (Autumn 1948), 86–88.

Cheng Z 1997
Cheng Zhijuan 程志娟. "The Han Guiju Mirror and *Liu Bo* 漢規矩鏡與六博." *Dongnan Wenhua* 東南文化 4 (1997), 124–26.

Chongqing 1986
Chongqing City Museum. "Chongqing shi lin jiang zhi lu xi han mu 重慶市臨江支路西漢墓." *Kaogu* 3 (1986), 230–42.

Chu 1935
Chu Hsi-Tsu. "The Tombs of the Six-Dynasties." *Monumenta Sinica* 1 (September 1935), 183–99.

Deng 1999
Deng Qiuling 鄧秋玲. "Zhan guo shan zi wen tong jing xi lun 戰國山字紋銅鏡析論." *Li shi wen wu yue kan* 歷史文物月刊 9 (3) (March 1999), 48–60.

Dong et al. 1962
Dong Gao 董誥 et al., eds. *Qin ding quan tang wen* 欽定全唐文. 18 vols. Taipei, 1962.

Farley 1940
Malcolm Farley. "Some Mirrors of Supposed Pre-Han Date." *Harvard Journal of Asiatic Studies* 5(1) (1940), 72–94.

Feng/Feng 1906
Feng Yunpeng 馮雲鵬 and Feng Yunyuan 馮雲鵷. *Jin shi shuo* 金石索. 24 vols. Shanghai, 1906.

Fu 1986
Fu Juyou 傅舉有. "On the Games, Game Articles and Cheeseboard Patterns on Bronze Mirrors of the Qin and Han Dynasties." *Kaogu Xuebao* 考古學報 1 (1986), 21–33.

Gansu/Zhangxian 1982
Gansu Provincial Museum and Cultural Centre of Zhangxian. "Excavation of the Yuan Dynasty Tombs of Family Wang Shixian at Zhangxian County in Gansu Province: Report 2." *Wenwu* 2 (1982), 13–21.

Gao 1991
Gao Zhixi. "On the Bronze Mirrors of Chu State." *Wenwu* 5 (1991), 42–60.

Ge 1929
Ge Hong 葛洪. *Xi jing za ji* 西京雜記. In *Si bu cong kan chu bian* 四部叢刊初編. Shanghai, 1929.

Guangdong/Dongguan 1991
Guangdong Provincial Museum and the Museum of the City of Dongguan. "Excavation of the Ming Tombs of Luo Hengxin Family in Dongguan, Guangdong." *Wenwu* 11 (1991), 38–50.

Guangdong/Huanan 1961
Guangdong Sheng Wenwu Guanli Weiyuan Hui 廣東省文物管理委員會 and Huanan Shifan Xueyuan Lishi Xi 華南師範學院歷史系. "Excavation of Zhang Jiuling's Tomb of the Tang Dynasty." *Wenwu* 6 (1961), 45–51.

Guangzhou 1981
CPAM of Guangzhou and the Municipal Museum of Guangzhou. *Excavation of the Han Tombs at Guangzhou.* 2 vols. Beijing, 1981.

Guizhou 1986
Guizhou Provincial Museum. "Excavation at Kele Township, Hezhang County, Guizhou Province 赫章可樂發掘報告." *Kaogu Xuebao* 2 (1986), 199–251.

Guo 1996
Guo Yuhai 郭玉海. *Gu gong cang jing* 故宮藏鏡. Beijing, 1996.

Haining 1983
Haining County Museum. "Zhejiang sheng hai ning xian dong shan song mu qing li jian bao 浙江海寧縣東山宋墓清理簡報." *Wenwu* 8 (1983), 25–32.

He T 1990
He Tangkun 何堂坤. "Alloy-composition Analysis of the Song Dynasty Bronze Mirror." *Sichuan wen wu* 四川文物 3 (1990), 74–79.

He T 1992
He Tangkun. *Zhong guo gu dai tong jing di ji shu yan jiu* 中國古代銅鏡的技術研究. Beijing, 1992.

He Z 1991
He Zhiguo, Mianyang Museum. "Excavation of the Eastern Han Cliff Tomb No. 1 at Hejiashan." *Wenwu* 3 (1991), 1–8.

Hebei 1990
Hebei Sheng Wenwu Yanjiu Suo 河北省文物研究所. "Report on the Excavation of Han Tombs at Beiguan, Yangyuan County, Hebei 河北陽原縣北關漢墓發掘簡報." *Kaogu* 4 (1990), 322–28.

Hebei 1996
Hebei Sheng Wenwu Yanjiu Suo. *Li dai tong jing wen shi* 歷代銅鏡紋飾 (Decorations on Ancient Chinese Bronze Mirrors). Shijiazhuang, 1996.

Hebei/Zhangjiakou 1990
Hebei Institute of Cultural Relics and Zhangjiakou Centre of Culture. "Hebei yang yuan san fen gou han mu qun fa jue bao gao 河北陽原三汾溝漢墓群發掘報告." *Wenwu* 1 (1990), 1–18.

Hefei 1991
CPAM, City of Hefei. "Northern Song Joint Burial Grave of Ma Shaoting and His Wife in Hefei." *Wenwu* 3 (1991), 26–38.

Hengyang 1986
Hengyang City Museum. "Excavation of Han to Southern Dynasties Tombs at Aodong, Chashan, Hengyang, Hunan Province 湖南衡陽茶山坳東漢至南朝墓的發掘." *Kaogu* 12 (1986), 1079–93.

Hengyang 1993
Hengyang Municipal Archaeological Team. "Excavation of Han Tombs at Fenghuangshan, Hengyang City, Hunan." *Kaogu* 3 (1993), 239–47.

Hepu 1995
Hepu County Museum. "Excavation of Han Tomb No. 10 at Fengmenling, Hepu County, Guangxi." *Kaogu* 3 (1995), 226–83.

Higuchi 1979
Higuchi Takayasu. *Kokyo.* 2 vols. Tokyo, 1979.

Huaiyin 1988
Huaiyin Museum. "A Report on the Excavation of Tomb No. 1 of Han Dynasty at Jiajiadun Site, Jiangsu Province." *Dong nan wen hua* 東南文化 1 (1988), 59–67.

Hubei/Echeng 1986
Hubei Provincial Museum and Echeng Municipal Museum. *Bronze Mirrors of the Han, Three Kingdoms, Six Dynasties Unearthed at Echeng* 鄂城出土漢三國六朝銅鏡. Beijing, 1986.

Hubei/Yunxian 1987
Hubei Provincial Museum and the Museum of Yunxian County. "Excavation of the Tang Tomb of Li Hui and Yan Wan at Yunxian, Hubei." *Wenwu* 8 (1987), 30–42.

Hucker 1985
Charles Hucker. *A Dictionary of Official Titles in Imperial China.* Stanford, 1985.

Hunan 1960
Hunan Provincial Museum. *Hunan chu tu tong jing tu lu* 湖南出土銅鏡圖錄. Beijing, 1960.

Hunan 1981
Hunan Provincial Museum. "Excavation of Eastern Han Tombs at Daoziping, Hengyang, Hunan Province." *Wenwu* 12 (1981), 35–37.

IOA 1959
Institute of Archaeology, Chinese Academy of Sciences. *Luo yang shao gou han mu* 洛陽燒溝漢墓. Beijing, 1959.

IOA 1966
Institute of Archaeology, Chinese Academy of Sciences. *Xi'an jiao qu sui tang mu* 西安郊區隋唐墓 (The Sui and Tang Tombs on the Outskirts of Xian). Beijing, 1966.

IOA 1980
Institute of Archaeology, Chinese Academy of Social Sciences. *Tang chang an cheng jiao sui tang mu* 唐長安城郊隋唐墓. Beijing, 1980.

Izumi-shi 1984
Izumi-shi Kobosō Kinen Bijutsukan. *Zō Kagami Takuei.* Izumi-shi, 1984.

Jiangxi/Xin'gan 1983
Jiangxi Sheng Wenwu Gongzuo Dui 江西省文物工作隊 and Xin'gan Xian Wenwu Chenglie Shi 新干縣文物陳列室. "Report on Western Jin Remains at Xin'gan, Jiangxi." *Kaogu* 12 (1983), 1122–24.

Jinan et al. 1984
Jinan Shi Wenguan Hui 濟南市文管會 et al. "The Dating and Other Problems for the Luohan Sculptures in Lingyan Temple, Changqing, Shandong Province 山東長清靈岩寺羅漢像的塑制年代及有關問題." *Wenwu* 3 (1984), 76–82.

Jingzhou 1995
Jingzhou Prefectural Museum. "Excavation of Western Han Tombs at Wafenyuan, Jianglin County, Hubei." *Kaogu* 11 (1995), 985–96.

Kerr 1990
Rose Kerr. *Later Chinese Bronzes.* London, 1990.

Kiangsi 1974
Kiangsi Provincial Museum. "The Tsin Tombs at Nanchang, Kiangsi Province." *Kaogu* 6 (1974), 373ff.

Kiangsu/Nanking 1964
CPAM, Kiangsu Province and the Nanking Museum. "Excavations of Han Tombs at San Yang Tun, Yen Ch'eng, Kiangsu Province." *Kaogu* 8 (1964), 394–402.

Kong 1980
Kong Xiangxing 孔祥星. "Sui tang tong jing di lei xing yu fen qi 隋唐銅鏡的類型與分期." *Zhong guo kao gu xue hui di yi ci nian hui lun wen ji* 中國考古學會第一次年會論文集. Beijing, 1980.

Kong 1992
Kong Xiangxing. *Zhong guo tong jing tu dian* 中國銅鏡圖典. Beijing, 1992.

Kong/Li 1992
Kong Xiangxing and Li Xuemei 李雪梅. "Guan yu jin dai tong jing shang di jian yan ke ji 關於金代銅鏡上的檢驗刻記." *Kaogu* 2 (1992), 170–74.

Kong/Liu 1994
Kong Xiangxing and Liu Yiman 劉一曼. *Zhong guo gu tong jing* 中國古銅鏡 (Ancient Chinese Bronze Mirrors). Taipei, 1994.

Lei 1982
Lei Congyun 雷從云. "Chu shi jing di lei xing yu fen qi 楚式鏡的類型與分期." *Jianghan Kaogu* 江漢考古 2 (1982), 20–36.

Li C 1957
Li Caiyu 李彩畲. "Cultural Relics from the White Pagoda, Congshou Temple, in Shenyang 瀋陽崇壽寺白塔的文物." *Kaogu Tongxun* 考古通訊 6 (1957), 34–38.

Li et al. 1959
Li Fang 李昉 et al., eds. *Tai ping guang ji* 太平廣記. Beijing, 1959.

Li ZG 1957
Li Zhenguang 李正光. "Lue tan chang sha chu tu di zhan guo shi dai tong jing 略談長沙的戰國時代銅鏡." *Kaogu Tongxun* 考古通訊 1 (1957), 96–106.

Liang 1940–42
Liang Shangchun 梁上椿. *Yan ku cang jing* 巖窟藏鏡. 6 vols. Beijing, 1940–42.

Lianyungang 1996
Museum of Lianyungang City. "Excavation of Some Han Tombs at Yinwan, Donghai, Jiangsu." *Wenwu* 8 (1996), 4–31.

Liaocheng 1993
Liaocheng Diqu Wenhuaju Wenwu Yanjiushi 聊城地區文化局文物研究室. "A Selection of Ancient Bronze Mirrors Unearthed from Liaocheng 聊城地區出土部分古代銅鏡." *Wenwu* 4 (1993), 90–93.

Liaoning 1975
Archaeological Team of the Liaoning Provincial Museum. "Excavation of the Liao Dynasty Tomb at Yehmaotai in Faku County, Liaoning Province." *Wenwu* 12 (1975), 26–36.

Liezi 1929
Liezi 列子. *Chong xu zhi de zhen jing*沖虛至德真經 In *Si bu cong kan chu bian* 四部叢刊初編. Shanghai, 1929?.

Liu A 1940
Liu An 劉安. *Huai nan hong lie jie* 淮南鴻烈解. In *Cong shu ji cheng* 叢書集成. Changsha, 1940.

Liu S 1997
Liu Shujuan 劉淑娟. *Liao dai tong jing yan jiu* 遼代銅鏡研究. Shenyang, 1997.

Liu T 1935
Liu Tizhi 劉體智. *Xiao jiao jing ge jin wen tuo ben* 小校經閣金文拓本. 1935.

Liu Y 1985
Liu Yiman. "Shi lun zhan guo tong jing di fen qu 試論戰國銅鏡的分區." *Kaogu* 11 (1985), 1008–14.

Lu 1978
Lu Jianguo 盧建國. "Exploratory Study of the Ruins of Yuhua Palace of the Tang Dynasty, at Tongchuan, Shaanxi 陝西銅川唐玉華宮遺址調查." *Kaogu* 6 (1978), 380–87.

Luo 1916
Luo Zhenyu 羅振玉. *Gu jing tu lu* 古鏡圖錄. 1916.

Luoyang 1982
The Archaeological Team of Luoyang City. "The Tomb of An Pu and His Wife of the Tang Dynasty at Longmen in Luoyang." *Zhong yuan wen wu* 中原文物 3 (1982), 21–26.

Luoyang 1988
Luoyang Museum. *Luoyang chu tu tong jing* 洛陽出土銅鏡. Beijing, 1988.

Luoyang 1992
Second Archaeological Team of Luoyang City. "Song Tomb with Fresco Paintings at Mangshan, Luoyang 洛陽邙山宋代壁畫墓." *Wenwu* 12 (1992), 37–51.

Luoyang 1996a
Second Archaeological Team of Luoyang City. "A Warring States Tomb to the North of Railroad in Luoyang." *Wenwu* 7 (1996), 39–41.

Luoyang 1996b
Second Archaeological Team of Luoyang City. "Excavation of a Tomb from the Wang Mang Period at Wunuzhong near Luoyang." *Wenwu* 7 (1996), 42–53.

Lüshun 1997
Lüshun Museum. *Lüshun bo wu guan cang tong jing* 旅順博物館藏銅鏡. Beijing, 1997.

Ma 1962
Ma Chengyuan 馬承源. "The Sword of the King of Yue, and the Mirror of Deities and Animals Dated to the First Year of Yonkang Era." *Wenwu* 12 (1962), 52–55.

Ma et al. 1998
Ma Chengyuan et al., eds. *Zhong guo qing tong qi quan ji* 中國青銅器全集. vol. 16. Beijing, 1998.

MacKenzie 1999
Colin MacKenzie. "The Influence of Textile Designs on Bronze, Lacquer and Ceramic Decorative Styles during the Warring States Period." *Orientations* (September 1999), 82–91.

Meeks 1993
Nigel Meeks. "Patination Phenomena on Roman and Chinese High-Tin Bronze Mirrors and Other Artefacts." *Metal Plating and Patination: Cultural, Technical, and Historical Developments.* Edited by Susan La Niece and Paul Craddock. Oxford, 1993.

Meyers/Holmes 1983
Pieter Meyers and Lore L. Holmes. "Technical Studies of Ancient Chinese Bronzes: Some Observations." *The Great Bronze Age of China: A Symposium.* Edited by George Kuwayamei. Seattle and London, 1983.

Mo 1958
Mo Yinghao. "Guang zhou hua qiao xin cun xi han mu 廣州華僑新村西漢墓." *Kaogu Xuebao* 2 (1958), 39–76.

Mongolia et al. 1996
Institute of Archaeology, Inner Mongolia et al. "Excavation of Yelu Yuzhi's Tomb of the Liao Dynasty." *Wenwu* 1 (1996), 4–32.

Nakano et al. 1994
Nakano Toru et al. *Bronze Mirrors from Ancient China: Donald H. Graham Jr. Collection.* Hong Kong, 1994.

National Palace 1989
National Palace Museum. *Gu gong shu hua tu lu* 故宮書畫圖錄. vol. 2. Taipei, 1989.

Nie 1994
Nie Shimei 聶世美. *Ling hua zhao ying* 菱花照影. Shanghai, 1994.

Ningxia 1978
Museum of Ningxia Huizu Autonomous Region. "Ningxia hui zu zi zhi qu wen wu kao gu gong zuo di zhu yao shou huo 寧夏回族自治區文物考古工作的主要收獲." *Wenwu* 8 (1978), 54–59.

O'Donoghue 1990
Diane M. O'Donoghue. "Reflection and Reception: The Origins of the Mirror in Bronze Age China." *The Museum of Far Eastern Antiquities Bulletin* 62 (1990), 5–183.

Omura 1915
Omura Seigai 大村西崖. *Shina Bijutsushi Chosohen* 支那美術史雕塑編. 2 vols. Tokyo, 1915.

Pei 1967
Pei Song 裴松. "San guo zhi 三國志: Wei zhi 魏志." In *Er shi si shi* 二十四史. Taipei, 1967.

Peng et al. 1985
Peng Zicheng 彭子成, Zhang Yanyao 張衍堯, and Liu Changfu 劉長福. "The Use of Ratios of Lead Isotope in Archaeological Research 鉛同位素比值法在考古研究中的應用." *Kaogu* 11 (1985), 1032–37.

Peng/Tang 1980
Peng Shifan 彭適凡 and Tang Changbo 唐昌朴. "Excavations of the Tombs of the Northern Song Dynasty in Jiangxi Province." *Wenwu* 5 (1980), 28–39.

Pingshuo 1988
Archaeological Team of Pingshuo, Shanxi. "Han Tomb No. 1 at Zhaoshibazhuang, Shuoxian, Shanxi 山西平朔趙十八莊一號漢墓." *Kaogu* 5 (1988), 442–48.

Quan 1998
Quan Hong. "On the Bronze Mirror of the Nanyue State." *Kaogu Xuebao* 3 (1998), 307–30.

Rupert/Todd 1935
Milan Rupert and O. J. Todd. *Chinese Bronze Mirrors.* Peiping, 1935.

Sanmenxia 1996
Sanmenxia Municipal Archaeological Team. "Excavation of Western Han Tombs at the Sanmenxia Municipal Thermal Power Plants, Henan." *Kaogu* 6 (1996), 6–15.

Santai 1976
Santai Xian Cultural Center 三台縣文化館. "Sichuan santai xian fa xian dong han mu 四川三台縣發現東漢墓." *Kaogu* 6 (1976), 395.

Shaanxi 1959
Shaanxi Sheng Wenwu Guanli Weiyuan Hui 陝西省文物管理委員會. *Shaanxi sheng chu tu tong jing* 陝西省出土銅鏡. Beijing, 1959.

Shōsōin 1994–97
Shōsōin Jimusho 正倉院事務所. *Shōsōin Hommotsu* 正倉院寶物. 7 vols. Tokyo, 1994–97.

Shunhua 1983
Shunhua Xian Wenhua Guan. "Shaanxi shun hua xian chu tu han dai tong jing 陝西淳化縣出土漢代銅鏡." *Kaogu* 9 (1983), 850–53.

Sichuan et al. 1989
CPAM, Sichuan Province, Culture Centre of the City of Mianyang and CPAM, Pingwu County. "Tombs of Wang Xi's Family of the Ming Dynasty at Pingwu." *Wenwu* 7 (1989), 1–42.

Sichuan/Chongqing 1960
Sichuan Provincial Museum and Chongqing City Museum. *Sichuan sheng chu tu tong jing* 四川省出土銅鏡. Beijing, 1960.

Sichuan/Qingchuan 1982
Sichuan Provincial Museum and Culture Centre, Qingchuan. "Wooden Tablets Inscribed with Revised Land Law of Qin—Excavation of the Warring States Tomb at Qingchuan, Sichuan Province." *Wenwu* 1 (1982), 1–13.

Siren 1956
Osvald Siren. *Chinese Painting: Leading Masters and Principles.* vol. III. New York, 1956.

Suzhou 1975
Suzhou Museum. "Report on the Excavation of Wang Xijue's Tomb 蘇州虎丘王錫爵墓清理記略." *Wenwu* 3 (1975), 51–56.

Taipei 1986
National Palace Museum. *Gu gong tong jing te zhan tu lu* 故宮銅鏡特展圖錄 (Catalogue of Special Exhibition of Bronze Mirrors in the National Palace Museum). Taipei, 1986.

Tang 1992
Tang Xianhua 唐先華. "A Tang Mirror with Twin Phoenixes Discovered at Hengyang, Hunan 湖南衡陽市發現唐代雙鳳鏡." *Kaogu* 11 (1992), 1054.

Taube et al. 1995
Michelle Taube, W. Thomas Chase, Alison J. Davenport, and A. Peter Jardine. "Investigation and Replication of the Surface Microstructure of Early Chinese Black Bronze Mirrors." *Materials Issues in Art and Archaeology IV.* Pittsburgh, 1995.

Teng/Shi 1993
Teng Yanzhen 滕延振 and Shi Shizhen 石世鎮. "A Mirror of 'Zhenzi Feishuang' from Haining, Zhejiang." *Wenwu* 2 (1993), 62.

Tengxian 1984
Tengxian Museum. "The Tomb of Su Yu of the Jin Dynasty at Tengxian, Shandong Province." *Kaogu* 4 (1984), 349–51.

Umehara 1943
Umehara Sueji 梅原末治. *Kan Sangoku Rikuchō Kinen Kyō Zusetsu* 漢三國六朝紀年鏡圖說. Kyoto, 1943.

Umehara 1984a
Umehara Sueji. *Shoko Kokyo Shuei* 紹興古鏡聚英 (Selected Ancient Mirrors Found at Shao-xing Tombs). Kyoto, 1984.

Umehara 1984b
Umehara Sueji. *Tokyo Taikan* 唐鏡大觀. Kyoto, 1984 (reprint of 1945 edition).

Wang E 1994
Eugene Yuejin Wang. "Mirror, Death, and Rhetoric: Reading Later Han Chinese Bronze Artifacts." *Art Bulletin* LXXVI-3 (September 1994), 511–34.

Wang F et al. 1969
Wang Fu 王黼 et al. *Ye zheng tang chong xiu xuan he bo gu tu lu* 亦政堂重修宣和博古圖錄. Taipei, 1969.

Wang SL 1987
Wang Shilun 王士倫. *Zhejiang chu tu tong jing* 浙江出土銅鏡. Beijing, 1987.

Wang Z 1988
Wang Zhengshu 王正書. "Shanghai fu quan shan xi han mu qun fa jue 上海福泉山西漢墓群發掘." *Kaogu* 8 (1988), 694–717.

Wang ZS 1981a
Wang Zhongshu 王仲殊. "Notes on the Triangular-rimmed Bronze Mirrors with Mythical Animal Designs Found in Japan." *Kaogu* 4 (1981), 346–58.

Wang ZS 1981b
Wang Zhongshu. "Takamatsuzuka Tomb in Japan: Inquiring about Its Date." *Kaogu* 3 (1981), 277–78.

Wang ZS 1982
Wang Zhongshu. "The Date of Takamatsuzuka Tomb and the Identity of the Deceased 關於日本高松塚古墳的年代和被葬者." *Kaogu* 4 (1982), 410–13.

Wang ZS 1984a
Wang Zhongshu. "A Synthetic Study on the Triangular-rimmed Bronze Mirrors with Mythical Figures and Animals Designs Found in Japan." *Kaogu* 4 (1984), 468–79.

Wang ZS 1984b
Wang Zhongshu. "Some Notes on Two Bronze Mirrors of the Cao-Wei Period." *Kaogu* 12 (1984), 1118–26.

Wang ZS 1985
Wang Zhongshu. "Wuxian, Shanyin and Wuchang: On the Producting [sic] Places of Bronze Mirrors of the Wu Kingdom as Indicated in Mirrors' Inscriptions." *Kaogu* 11 (1985), 1025–31.

Wang ZS 1986
Wang Zhongshu. "Some Notes on the Bronze Mirrors with Divinity and Animal Designs Attributed to Chen Shi, a Mirror Maker of Wu." *Kaogu* 11 (1986), 1017–25.

Wang ZS 1994
Wang Zhongshu. "On the Bronze Mirror with the Dated Inscription '3rd Year, Qinglong Reign' (Wei Dynasty of Ancient China) Unearthed in Japan." *Kaogu* 8 (1994), 727–35.

Wang/Li 1987
Wang Yulang 王禹浪 and Li Chenqi 李臣奇. "Jin dai tong jing chu bu yan jiu 金代銅鏡初步研究." In Chen Shu 陳述, *Liao jin shi lun ji* 遼金史論集. vol. III. Beijing, 1987.

Wen 1956
Wen You 聞宥. *Sichuan han dai hua xiang xuan ji* 四川漢代畫像選集. Beijing, 1956.

Wirgin 1970
Jan Wirgin. *Sung Ceramic Designs*. London, 1970.

Wu H 1989
Wu Hong. *The Wu Liang Shrine: The Ideology of Early Chinese Pictorial Art*. Stanford, 1989.

Wu S 1993
Wu Shuicun 吳水存. *Jiujiang chu tu tong jing* 九江出土銅鏡. Beijing, 1993.

Wu W 1991
Wu Wei 吳煒. "Jiangsu yi zheng xu pu fa xian tang mu 江蘇儀徵胥浦發現唐墓." *Kaogu* 2 (1991), 188–90.

Xianning/Yangxin 1994
Xianning Prefectural Museum and Yangxin County Museum. "Tomb 1 of the Warring States Period at Banbishan, Yangxin, Hubei." *Kaogu* 6 (1994), 525–31, 537.

Xiong 1979
Xiong Chuanxin 熊傳新. "On the *Lubo* Game from the Mawangdui No. 3 Han Tomb." *Wenwu* 4 (1979), 35–39.

Xu D 1994
Xu Diankui 徐殿魁. "An Archaeological Study of the Periodization of the Tang Mirror." *Kaogu Xuebao* 3 (1994), 263–342.

Xu N 1914
Xu Naichang 徐乃昌. *Xiao tan luan shi jing ying* 小檀欒室鏡影. 18 vols. 1914.

Xu Y 1986
Xu Yongxiang 徐勇翔. "Study of the Cloud-Dragon Pattern in Jade Ornamentation of the Tang Period 唐代玉雕中的雲龍紋裝飾研究." *Wenwu* 9 (1986), 58–63.

Yang 1994
Yang Guirong 楊桂榮. "Guan cang tong jing xuan ji, Part 6." *Zhong guo li shi bo wu guan guan kan* 中國歷史博物館館刊 2 (1994), 126–44.

Yang 1995a
Yang Guirong. "Guan cang tong jing xian ji, Part 7." *Zhong guo li shi bo wu guan guan kan* 1 (1995), 135–43.

Yang 1995b
Yang Guirong. "Guan cang tong jing xuan ji, Part 8." *Zhong guo li shi bo wu guan guan kan* 2 (1995), 111–20.

Yangzhou/Hanjiang 1991
Yangzhou Museum and Hanjiang County Library. "Jiangsu han jiang xian yang shou xiang bao nu dun xin mang mu 江蘇邗江縣楊壽鄉寶女墩新莽墓." *Wenwu* 10 (1991), 39–61.

Yanshi 1986
Second Archaeological Team, IA, CASS. "Excavation of Six Dated Tombs of the Tang Dynasty at Xingyuancun, Yanshi, Henan." *Kaogu* 5 (1986), 429–62.

Yanshi 1996
Second Henan Archaeological Team, IA, CASS. "Excavation of Tang Tombs at Xinyuan Village, Yanshi County, Henan." *Kaogu* 12 (1996), 1–35.

Yen 1989
Yen Chuan-ying 顏娟英. "Tang dai tong jing wen shi zhi nei rong yu feng ge 唐代銅鏡文飾之內容與風格." *Zhong yang yan jiu yuan li shi yu yan yan jiu suo ji kan* 中央研究院歷史語言研究所集刊 60-2 (June 1989), 289–366.

Yichang/Yidu 1988
Yichang Regional Museum and Yidu Cultural Institute. "Relics from the Eastern Han Burials at Lucheng, Yidu." *Kaogu* 10 (1988), 910–17.

Yiyang 1981
Yiyang Cultural Institute. "Tang Tombs at Heshan Temple, Yiyang, Henan." *Kaogu* 4 (1981), 315–18.

Yu 1958
Yu Linxiang 于臨祥. "Yin cheng zi bei mu 營城子貝墓." *Kaogu Xuebao* 4 (1958), 71–90.

Yuen 1986
Yuen Ting Cheuk. "A Study of Hu Chou Mirrors in Southern Sung Dynasty." *Journal of the Institute of Chinese Studies of the Chinese University of Hong Kong* XVII (1986), 145–61.

Zhang J 1982
Zhang Jinyi 張金儀. *Han jing suo fan ying de shen hua chuan shuo yu shen xian si xiang* 漢鏡所反映的神話傳說與神仙思想. Taipei, 1982.

Zhang Q 1992
Zhang Qiming 張啟明. "Selected Bronze Mirrors from the Langzhong Collection." *Sichuan wen wu* 四川文物 6 (1992), 21–24.

Zhang T et al. 1967
Zhang Tingyu 張廷玉 et al., eds. *Ming shi* 明史. In Baina edition of *Er shi si shi* 二十四史. Taipei, 1967.

Zhang Y 1990
Zhang Ying 張英. *Jilin chu tu tong jing* 吉林出土銅鏡 (Bronze Mirrors Excavated in Jilin). Beijing, 1990.

Zhang Y et al. 1967
Zhang Yuanji 張元濟 et al., eds. *Er shi si shi* 二十四史. 41 vols. Taipei, 1967.

Zhejiang 1988
Zhejiang Institute of Archaeology. "Song Tombs at Beidaqiao, Hangzhou 杭州北大橋宋墓." *Wenwu* 11 (1988), 55–60.

Zheng 1940
Zheng Qi 鄭綮. *Kai tian chuan xin ji* 開天傳信記. In *Cong shu ji cheng chu bian* 叢書集成初編. Changsha, 1940.

Zhengzhou 1995
Zhengzhou Shi Wenwu Gongzuo Dui 鄭州市文物工作隊. "Tang Tombs Found in Zhengzhou 鄭州地區發現的幾座唐墓." *Wenwu* 5 (1995), 23–39.

Zhenjiang 1984
Museum of Zhenjiang. "Excavations of Some Tombs of Eastern Wu and Western Jin Periods in Zhenjiang." *Kaogu* 6 (1984), 528–45.

Zhonghua 1960
Zhonghua Shuju 中華書局, ed. *Quan tang shi*. 12 vols. Beijing, 1960.

Zhou SR 1986a
Zhou Shirong 周世榮. "Hunan chu tu han dai tong jing wen zi yan jiu 湖南出土漢代銅鏡文字研究." *Gu wen zi yan jiu* 古文字研究 XIV (1986), 69–175.

Zhou SR 1986b
Zhou Shirong. *Tong jing tu an—hunan chu tu li dai tong jing* 銅鏡圖案—湖南出土歷代銅鏡. Changsha, 1986.

Zhou Z 1987
Zhou Zheng 周錚. "'Gui ju jing' ying gai cheng 'bo ju jing' 規矩鏡應改稱博局鏡." *Kaogu* 12 (1987), 1116–18.

Zhou/Zhou 1979
Zhou Xin 周欣 and Zhou Changyuan 周長源. "Tang Mirrors Unearthed in Yangzhou 揚州出土的唐代銅鏡." *Wenwu* 7 (1979), 53–58.

Zhu 1988
Zhu Zhi 朱幟. "Bronze Mirror Unearthed from Wuyang, Henan 河南舞陽出土戰國銅鏡." *Kaogu* 5 (1988), 472.

Zhu/He 1993
Zhu Shoukang and He Tangkun. "Studies of Ancient Chinese Mirrors and Other Bronze Artefacts." *Metal Plating and Patination: Cultural, Technical, and Historical Developments*. Edited by Susan La Niece and Paul Craddock. Oxford, 1993.

Zhucheng 1987
Zhucheng County Museum. "Shandong zhu cheng xian xi han mu guo mu 山東諸城縣西漢木槨墓." *Kaogu* 9 (1987), 778–85.